Techniques of
SURFACE AND COLLOID
CHEMISTRY AND PHYSICS

VOLUME 1

EDITED BY

R. J. GOOD
Department of Chemical Engineering
State University of New York at Buffalo
Buffalo, New York

R. R. STROMBERG
Institute for Materials Research
National Bureau of Standards
Washington, D.C.

R. L. PATRICK
Alpha Research & Development, Inc.
Elverson, Pennsylvania

MARCEL DEKKER, INC. New York   1972

CHEMISTRY

MARCEL DEKKER, INC.
95 Madison Avenue, New York, New York 10016

LIBRARY OF CONGRESS CATALOG CARD NUMBER 71-184201

ISBN 0-8247-1262-5

PRINTED IN THE UNITED STATES OF AMERICA

PREFACE

Measurement, as is well known, is the essence of experimental physical chemistry and physics. A very large number of the techniques that have led to the major advances in the science of colloids and surfaces are methods of measurement of physical properties or of changes that occur in the systems.

Of the experimental techniques that are available, there are a number that are specific to problems in surface and colloid chemistry and physics. There are other techniques that have more general applicability, but require special refinements or interpretations when applied to surfaces or colloids. Finally, of course, there are those techniques which are useful for a wide range of problems and require no additional refinements when used for studies of surfaces or colloids.

Together, all these provide a wide spectrum of techniques for obtaining information useful for the characterization of surfaces and colloids and their interactions and reactions.

A scientist seldom has expertise in all the techniques he might like to use. The purpose of this series is to provide a compendium of the important techniques in the field, each described in sufficient detail to be immediately useful. This means that the fundamental concepts for each property and its measurement must be presented, the latest equipment and procedure described in appropriate detail, the computations explained, the limits of accuracy indicated, and the pitfalls pointed out. When more than one method for studying a property or process is available, it is important that the reader should be able to make a selection. Accordingly, the authors and editors must use their best scientific judgement and experience, to indicate which method is "best", and under what circumstances.

The present volume includes two topics that are closely related to each other: the film balance, discussed by N. L. Gershfeld, and monolayer permeability, discussed by M. Blank. The topics of bilayer lipid membranes, covered by Tien and Howard, and of ultrafiltration, covered by van Oss, have more apparent than substantial features in common. For example, Chapter 4 covers the numerous other properties of bilayer membranes, in addition to permeability. The subject of Chapter 5, the visualization of concentration gradients, by van Oss, is of great importance to the study of colloidal systems; it is of great indirect interest to surface science in that other phenomena, such as surface charge or membrane permeability, can be studied by means of transport phenomena; and this calls for the quantitative observation of concentration gradients.

The intended scope of this series is, the entire field of colloid and surface chemistry and physics. This means, all types of dispersed systems and all types of surfaces and interfaces, and the processes and interactions that are important in such systems. The authors and topics in future volumes are being chosen with this objective; and subsequent volumes should come out at approximately one year intervals.

Robert J. Good
R. R. Stromberg
R. L. Patrick

CONTRIBUTORS TO VOLUME I

Martin Blank, Department of Physiology, Columbia University, College of Physicians and Surgeons, New York, New York 10032

N. L. Gershfeld, Laboratory of Physical Biology, National Institute of Arthritis and Metabolic Diseases, National Institutes of Health, Bethesda, Maryland

Robert E. Howard, Department of Pathology, Medical School at San Antonio, The University of Texas, San Antonio, Texas

Carel J. van Oss, Department of Microbiology, School of Medicine, State University of New York at Buffalo, Buffalo, New York

H. Ti Tien, Department of Biophysics, Michigan State University, East Lansing, Michigan

CONTENTS

Chapter I

FILM BALANCE AND THE EVALUATION OF INTERMOLECULAR
ENERGIES IN MONOLAYERS

N. L. Gershfeld

Laboratory of Physical Biology
National Institute of Arthritis and Metabolic Diseases
National Institutes of Health
Bethesda, Maryland

I. INTRODUCTION

The film balance is the principal instrument for studying the thermodynamic properties of insoluble films on liquid surfaces. The most commonly used models can be traced back to Pockels [1], who first introduced
the use of the trough, barriers, and float for manipulating the insoluble
films. Subsequent innovations were made by Langmuir [2], who developed
the horizontal float method for measuring the surface pressure directly, and
later by Adam and Jessop [3], who perfected the end-loop system for
attaching the float to the sides of the trough while maintaining the necessary
float mobility. The Wilhelmy plate [4-6], which along with the float system

1

represents the most commonly used sensing device for surface pressure, is very useful for studying the properties of liquid–liquid interfaces.

Physical and chemical insights into the nature of insoluble monolayers generally followed a parallel development with instrumentation. Thus, the monomolecularity of oil films on water [7], the importance of molecular shape and the orientation of the molecules at the interface [2,8], and the existence of discrete monolayer states analogous to those physical states which exist in bulk [9] — concepts which have helped shape contemporary ideas on the structure of surfaces — were demonstrated with the film balance or one of its prototypes.

The quantitative treatment of the molecular properties of films, however, has been less than satisfactory. There is, for example, no adequate interpretation of the various phase transitions that have been observed with condensed lipid monolayers on water. The lack of quantization is due partially to the absence of a sufficiently detailed thermodynamic model of the system. For lipid films on water this would entail knowing the configuration of the hydrocarbon chain and the mutual van der Waals interaction energy, the polar group contribution to total energy, and, for aqueous substrates, the contribution of the hydration shell to the properties of the film as a function of the surface concentration of the lipid.

In the past, a good deal of attention was focused on the problems of reproducibility and elimination of experimental artifacts. These aspects of the film balance have been treated extensively elsewhere [10] and do not need repetition except in certain details. However, the experimental techniques for evaluating intermolecular energies in monolayers have not been systematically examined. In this chapter I shall outline some of the experiments for obtaining a thermodynamic model of insoluble monolayers on liquid surfaces. Of the two sections which follow, the first considers the principles upon which the experimental approach was developed; the second section presents experimental details. The appendix at the conclusion of this chapter covers aspects of the kinetics of monolayer desorption.

The major emphasis will be on the properties of lipid monolayers, since they are perhaps the most accessible experimentally of all the monolayer systems which have been studied. However, the experimental approaches discussed are generally applicable to other systems. This chapter is not intended as a review of the literature of the properties of monolayers on fluids. By way of justifying omission of much of the surface chemistry literature, it is noted that only a small proportion of the published data has dealt directly with the problems to be discussed here.

II. INTERMOLECULAR ENERGIES IN LIPID MONOLAYERS

The most general approach for obtaining the intermolecular energies in lipid films is to examine the sequence of physical processes diagramed in Fig. 1. (AB) is the process of adsorption of lipid molecules from the (bulk)

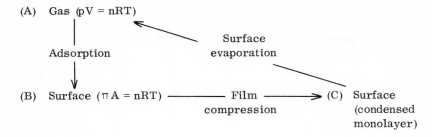

Fig. 1. Cyclical process for evaluation of intermolecular energies in monomolecular films. (A) is the (bulk) ideal gas state of film material; (B) and (C) are the ideal gas and condensed surface states, respectively.

gaseous state to the surface gaseous state; in both the initial and final states of this process, the lipid molecules exhibit ideal gas behavior, i.e., there are no lipid–lipid interactions. Defined in this way, the adsorption process involves hydration (in the case of the water surface) and possible changes in molecular configuration of the adsorbed lipids. Process (BC) is film compression — the transformation of the monolayer from the ideal gas to a condensed film — and involves changes in lipid–lipid interactions as well as possible changes in film hydration. Process (CA) is surface evaporation — the transfer of lipid molecules from a condensed surface state to the bulk gaseous state. The surface evaporation process is a function of all intermolecular energies in the condensed monolayer. Evaluation of the energies of any two of the processes automatically gives the third.

The cyclical process outlined in Fig. 1 serves to emphasize the difficulty of determining the contribution of the substrate to the internal energy of the film. Fig. 1 indicates that solvation of the monolayer can only be obtained from either the adsorption process or the surface evaporation process. To evaluate the energetics of adsorption and surface evaporation, the properties of the bulk vapor state of the film-forming compounds must be known. Unfortunately, this information is presently unavailable. Thus, until the bulk vapor state of the compounds can be understood, the solvation energies of the monolayers cannot be treated rigorously.

The following section considers the process of film compression in detail. For the sake of completeness, general methods for obtaining the energies for adsorption and surface evaporation will be cursorily outlined, despite the fact that they require methods outside the scope of this chapter.

A. Film Compression

There are two approaches for evaluating the changes in energies in this process. One method — the more general of the two — is to use the surface pressure-area isotherms directly, and the other is to use desorption experiments. The equivalence of the methods has been demonstrated [11]. We shall examine the former method first.

1. The π-A Isotherms of a Single Insoluble Component

The surface pressure π is defined by the relation

$$\pi = -\left(\frac{\partial F}{\partial A}\right)_T \tag{1}$$

where F is the Helmholtz free energy of the system and A is the area of the surface, with the temperature (T), total volume, and quantities of all components in the entire system held constant. Integration of Eq. (1) between the limits of A and A_i yields

$$F - F_i = -\int_{A_i}^{A} \pi dA \tag{2}$$

If A_i is chosen such that $\pi A_i = kT$, i.e., where the insoluble monolayer behaves as a two-dimensional ideal gas, then $F - F_i = \Delta F_c$, where ΔF_c is the Helmholtz free energy of film compression. The choice of $\pi A_i = kT$ as a reference state in effect defines ΔF_c as the work of compressing the insoluble monolayer from a state in which no intermolecular film contacts occur (the "ideal gas" region) to some condensed state where intermolecular contacts result in the repulsive and cohesive energies which are characteristic of the state. *

Evaluation of ΔF_c may be obtained directly if the equation of state is known. However, for most condensed monolayer systems the equation of state is not available and ΔF_c must be obtained by graphical integration of the isotherm. This entails measuring the very low surface pressures in the region where the film approaches ideal gas behavior [11]. The great majority of film balance experiments in the literature have been limited to a very narrow range of surface concentrations — mainly the condensed regions of the isotherms. As indicated earlier, these limited studies have been useful primarily in the development of a qualitative picture of the physical state of the film and the shapes of the molecules in the surface. However, for a condensed monolayer, such as stearic acid on water, calculation of ΔF_c for only this limited region of the isotherm ignores perhaps as much as 90% of the work of compressing the monolayer from the ideal gas to the condensed state. From the standpoint of evaluating energies of intermolecular interaction this is a very serious omission.

———

*It is perhaps also useful to characterize the ideal gas as a two-dimensional surface solution in which the surface pressure arises from osmotic pressure difference between an interface containing a second component (the insoluble film) and a surface of pure solvent [12,13]. The virtue of this model is that it explicitly treats the solvent as a component of the surface film. The equation assumes the same form for the two-dimensional ideal solution model as the ideal gas model. With respect to film compression, it is obviously immaterial which model one chooses for the reference state — the energy for the process of film compression will be identical.

Figure 2 illustrates the type of isotherm for condensed lipid films on water. Several features are noteworthy. The horizontal portion of the isotherm is the transition region between gaseous and condensed states. The constant value of the surface pressure in the transition region is π_v, the surface vapor pressure. A_i, A_v, and A_c are the areas at which the film behaves as an ideal gas, the vapor-gas transition area, and the condensed film area, respectively. For normal aliphatic monolayers which form solid, liquid-condensed, or liquid-expanded films, A_c is ~20, ~25, or 40-50 Å^2 per molecule, respectively. A_v may be of the order of 1000-5000 Å^2 per molecule, while A_i is of the order of 50,000 Å^2 per molecule [11].

Methods for obtaining the π-A isotherms in the region of A_i, and hence ΔF_c, are given in the experimental section of this chapter. The companion enthalpy and entropy of compression, ΔH_c and ΔS_c, may be obtained from the temperature dependence of ΔF_c.

2. ΔF_c from the Free Energy of Desorption

For an adsorbed film in equilibrium with a solution of surface-active molecules, the standard free energy of desorption, i.e., the transfer of surface molecules to the adjoining bulk phase, may be written as

$$\lambda = RT \ln \frac{\Gamma}{\delta C} \frac{\gamma^*}{\gamma} \tag{3}$$

where Γ is the surface concentration in moles per square centimeter, C is the bulk concentration of the surface-active compounds in moles per cubic centimeter, γ^* is the monolayer activity coefficient, γ is the activity coefficient in bulk solution, and δ is the insoluble film thickness. Moreover,

$$\lim_{\Gamma \to 0} \gamma^* = 1, \text{ and } \lim_{C \to 0} \gamma = 1$$

Implicit in these definitions is that γ^* takes into account all interactions in the monolayers. It is readily shown by comparing ΔF_c with the work of compressing an ideal gas (i.e., $\pi A = kT$) that an alternative expression for γ^* is

$$\gamma^* = \frac{A}{A_i} \exp \frac{F-F_i}{RT} \tag{4}$$

where A, A_i, F, and F_i are defined in Eq. (2). Thus, from Eq. (3), by measuring Γ/C for the ideal gas films (i.e., when $\gamma^*=1$), and for the condensed films, $F - F_i = \Delta F_c$ may be obtained. In principle, these relations may be applied to all monolayer systems since no monolayer is completely insoluble in the substrate.

For compounds with long hydrocarbon "tails" which are only very slightly soluble in water, measurement of Γ/C, and hence ΔF_c, is extremely difficult. However, it is possible to obviate the measurement of Γ/C by measuring instead the rates of monolayer desorption [14-16]. One can readily visualize how the rates of desorption are related to the energy of the film molecules since the higher the potential energy of the film molecules, the easier it is for them to escape into the adjoining bulk phase. This section concludes with an outline of the kinetics of monolayer desorption and its formal relation to ΔF_c.

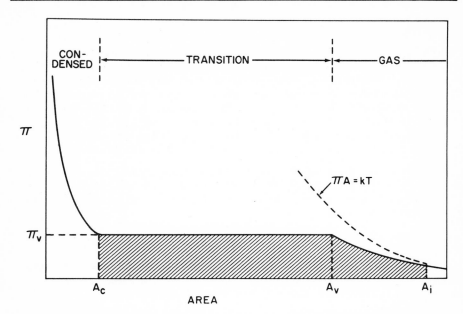

Fig. 2. Schematic representation of π-A isotherm in the transition region between a condensed state, area A_c, and a gas, area A_v. The surface vapor pressure, π_v, is given by the value of the horizontal line in the diagram. The magnitude of the shaded area is ΔF_c (see text). Reproduced from Ref. [11] by permission.

If the slightly soluble surface-active compound is spread as a monolayer, the film is unstable and desorbs from the surface. (In this sense desorption can also include evaporation from the surface to the bulk phase vapor state.) A model for desorption which is applicable to many monolayer systems [14-16] is schematically represented in Fig. 3. The model is based on the assumption that the rate of desorption is influenced by the rate of dissolution (i. e., the net transfer of film molecules from the surface to a very thin region α just beneath the film) and by the rate of diffusion of dissolved film away from the surface through an unstirred region of thickness ε to the bulk of the solution. In addition, it is assumed that the concentration of dissolved film molecules in the bulk of the solution is essentially zero.

According to this model, when the dissolution process is much faster than diffusion away from the surface, dissolved film molecules accumulate in region α (Fig. 3); a condition of virtual equilibrium is established between region α and the surface film from the very beginning of the desorption experiment. The concentration C of dissolved film molecules in region α will be determined by the Gibbs adsorption isotherm and hence will depend on Γ and π.

Diffusion proceeds from region α across the unstirred layer ε, where film molecules are removed to the bulk of the solution by convection

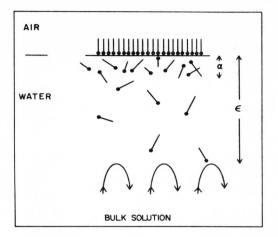

Fig. 3. A schematic representation of the physical model used for film desorption. For description see text. Reproduced from Ref. [16] by permission.

currents. After a short while a steady state is established where the rate of film loss equals the rate of solute diffusion.

If A_T is the total area of the film at time t and surface pressure π, the rate constant for desorption in the steady state k^{eq} is defined by

$$k^{eq} = -\left(\frac{d \ln A_T}{dt}\right)_\pi \tag{5}$$

where superscript eq is used to indicate the virtual equilibrium between the surface film and dissolved solute. Solution of the diffusion equation for this model [14] leads to the relation

$$k^{eq} = \frac{D C}{\epsilon \Gamma} \tag{6}$$

where D is the solute diffusion coefficient; for small C, D should be constant. From Eqs. (3), (4), and (6), assuming the $\gamma = 1$, eliminating Γ/C from the equations,

$$k^{eq} = \frac{D}{\epsilon \delta} \frac{A}{A_i} \exp (\Delta F_c - \lambda)/RT \tag{7}$$

In Eq. (7) ϵ/D may be evaluated from the desorption kinetics [14, 16] and δ may be estimated from the extended length of the hydrocarbon chain [17]. The standard free energy of desorption λ may be estimated from the surface tensions of dilute solutions of the lower, more soluble homologues of the "insoluble" films [18-20]. Thus, ΔF_c for long chain aliphatic compounds also can be evaluated from desorption experiments.

A simple test for this model of film desorption is the surface pressure dependence of k^{eq} given by the equation

$$\frac{d \ln k^{eq}}{d\pi} = \frac{A}{RT} - C_s \qquad\qquad (8)$$

where A is the film area per molecule, R is the gas constant, T is the absolute temperature, and C_s is the surface compressibility [15].

However, this model for desorption is not expected to be generally applicable, especially since it is unlikely that the activation energy for the dissolution process will always be very low relative to that for diffusion. Indeed, Eq. (8) has not been found to hold in all cases of film desorption [15]. A more detailed and general analysis of the kinetics of desorption [16] takes into account that film dissolution may also limit the kinetics of monolayer desorption. The results of this treatment take the form of a simple correction term for Eq. (8). Details of the correction term and methods for its evaluation are given in the Appendix.

B. Surface Evaporation and Adsorption

The processes of surface evaporation and adsorption are treated together in this section. They are defined arbitrarily to emphasize that surface evaporation is a measure of the energy of the condensed monolayer, while adsorption is the process to be used for evaluating the energies in the gaseous (monolayer) state. For the latter, the free energy of adsorption is λ [see Eq. (3)].

As indicated earlier, these processes are essential for evaluating the contribution of the substrate to the energy of the film. An outline of methods for treating surface evaporation and adsorption follows.

There are two methods which may be applied to measuring the heats of surface evaporation. The first of these follows directly from the observation that long chain aliphatic alcohols evaporate from water surfaces [21-23]. In principle, the rates of surface evaporation may be treated formally in the same way as the rates of monolayer desorption discussed in the previous section. Available data for rates of surface evaporation indicate that the alcohols obey Eq. (8); therefore, these systems are diffusion controlled [15]. Thus, for surface evaporation, an equation similar in form to Eq. (7), but relating k^{eq} to ΔF_v (the free energy of surface evaporation), can be derived. The heat of evaporation ΔH_v is obtained from $d\Delta F_v/dT$.

The second method assumes that the heat of surface evaporation of the alcohol is equal to the difference between the heat of sublimation (or the heat of vaporization) of the pure solid (or liquid) bulk phase of the film-forming compound, and its heat of spreading [24]. The latter may be obtained from the temperature dependence of the equilibrium spreading pressures of the compounds [25].

The adsorption energies depend on obtaining ideal gas properties of the vapor (bulk) state of the film-forming compounds. Since many of the film-forming compounds have extremely low vapor pressures, only estimates of these properties can presently be obtained. Estimates of the energies of adsorption for the lower, more volatile homologues of these compounds have been obtained [18] and should be consulted for additional insights into the problem of measuring heats of adsorption at water surfaces.

C. Properties of Mixed Monolayers

When two or more components are present in the surface, the system is referred to as a mixed monolayer. In principle, one may treat the surface mixing process as a bulk solution process. However, unlike the situation in bulk solution processes, it is difficult to detect immiscibility, i.e., no mixing in the monolayer. In this section we shall first consider methods for ascertaining whether the components are miscible, and then examine procedures for obtaining excess thermodynamic properties of mixing in monolayers.

1. Methods for Establishing Miscibility in Monolayers

a. Gibbs Phase Rule. Defay [26a] and more recently Crisp [26b] have extended the phase rule of Gibbs to include surfaces explicitly. For variable surface tension (or surface pressure), temperature, and external pressure

$$F = C^B + C^S - P^B - P^S + 3 \tag{9}$$

where F is the number of degrees of freedom, C^B is the number of components in bulk which are equilibrated throughout, C^S is the number of components restricted to the surface, P^B is the number of bulk phases, and P^S is the number of monolayer phases in equilibrium with each other. To illustrate the use of the phase rule we shall examine several probable situations.

To test whether two surface components at the air-water interface, each in a condensed monolayer state (e.g., liquid-condensed or liquid-expanded), are mutually miscible, mixing of the two components is done at a surface concentration where each component is in equilibrium with its own surface vapor, i.e., in the transition region of the isotherm (see Fig. 2). In this case $C^B = 2$ (air and water), $C^S = 2$ (components A and B), and $P^B = 2$ (air and water). The phase rule is then written as

$$F = 5 - P^S \tag{10}$$

If A and B are miscible, $P^S = 2$ (condensed film AB, and vapor AB) and F = 3. For constant temperature and pressure, there will be one degree of freedom, and π will vary with the composition of the surface. If A and B are immiscible, $P^S = 3$ (the two condensed phases of A and B, plus the surface vapor); for constant temperature and pressure, F = 0, and the surface pressure will be independent of the composition of the surface.

Figure 4 illustrates the type of results one may expect, where the surface vapor pressure π_v is plotted as a function of X_A, the mole fraction of A in the surface. Curve 1 is for complete immiscibility, where the measured surface pressure will be $(\pi_v{}^A + \pi_v{}^B)$. Curve 2 represents partial miscibility, where the region between points a and b represents the coexistence of two condensed phases with mole fraction composition $X_A{}^a$, $X_A{}^b$. Complete miscibility will be noted by curves 3 and 4, which vary continuously as a function of X_A between the two vapor pressures. The dotted line indicates ideal behavior, i. e., where the monolayer analogue of Raoult's law is obeyed.

b. <u>Film Penetration.</u> For a two-component surface film the Gibbs adsorption equation is written

$$d\pi = \Gamma_A \, d\mu_A + \Gamma_B \, d\mu_B \tag{11}$$

where Γ is the surface excess concentration of the component, and μ is the chemical potential. If one of the components (e. g., B) is soluble in the bulk

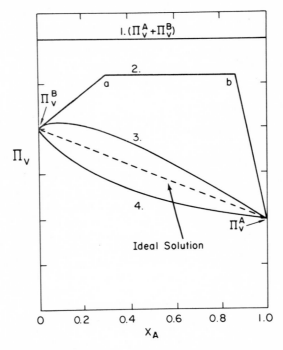

Fig. 4. Surface vapor pressure π_v vs. mole fraction X_A for surface solution containing two insoluble components A and B. Curve 1, complete immiscibility; curve 2, partial miscibility; curves 3 and 4, complete miscibility; dotted line indicates ideal solution behavior, i. e., the surface solution obeys Raoult's law.

solution phase, film penetration experiments can be used to test for surface miscibility directly. If insoluble monolayer A is spread on a solution of B, miscibility of B in A will be manifested as a change in π. If spreading of A occurs directly from a bulk phase (i. e., liquid or crystal), the equilibrium spreading pressure π^e is monitored. In this case $d\mu_A = 0$, and the Gibbs equation may be integrated to yield

$$\Delta \pi^e = \int \Gamma_B \, RT \, d \ln C_B \qquad (12)$$

By plotting π^e as a function of concentration of B in the bulk solution C_B, the slope of the curve will give Γ_B. Alternatively, radiotracers of B may be used to measure Γ_B directly (see Sec. III).

2. Excess Thermodynamic Properties of Mixing

An excess thermodynamic function is defined as the difference between the value of the function in a given mixture and in an ideal mixture of the same composition. Since Goodrich [27] first indicated the utility of evaluating the excess free energy of mixing for describing interactions between two insoluble film components, considerable efforts have been devoted to the measurements of these energies.

The excess Gibbs free energy for the mixing of two insoluble film components A and B may be written as

$$\Delta G_{mix}^{excess} = \int_0^\pi \sigma_{AB} \, d\pi - N_A \int_0^\pi \sigma_A \, d\pi - N_B \int_0^\pi \sigma_B \, d\pi \qquad (13)$$

where N_A and N_B are the total number of moles of A and B, and $N_A + N_B = 1$; σ_A and σ_B are the molar areas of the two pure films; and σ_{AB} is the mean molar area in the mixed film, i. e.,

$$\sigma_{AB} = N_A \bar{\sigma}_A + N_B \bar{\sigma}_B \qquad (14)$$

where $\bar{\sigma}_A$ and $\bar{\sigma}_B$ are the partial molar areas in the mixture. Each of the integrals in Eq. (13) may be evaluated from the π–A isotherms of the pure and mixed films.

In the original formulation of the excess free energies of mixing, it was assumed that as $\pi \to 0$, $\Delta G_{mix}^{excess} = 0$. However, for many of the studies which have been reported, this assumption is not valid. Due to the relative insensitivity of typical surface pressure measurements, values of π taken to be zero often fall in the region of the π–A isotherm where ΔG_{mix}^{excess} is not zero. This is particularly true when one of the surface components forms a condensed film (e. g., liquid-condensed or liquid-expanded) by itself. For these condensed systems, the condition of $\pi = 0$ is taken mistakenly at the point of the isotherm equivalent to A_c (see Fig. 2); graphical integration of

Eq. (13) often has ignored the portion of the isotherm to the right of A_c. Fig. 2. The error in $\Delta G \, _{\text{mix}}^{\text{excess}}$ which is introduced by ignoring the latter portion of the π-A isotherm may be of the same order of magnitude as the excess mixing energies themselves. To avoid this error, it is necessary to measure the isotherms at least into the region where the mixed film is gaseous (Fig. 2). In the experimental section which follows, a method is discussed for measuring the low surface pressures necessary to fulfill this condition.

Given the proper measurements of π, the procedure for measuring the excess thermodynamic properties of mixing should be to first establish the limits of miscibility, and second, to evaluate the π-A isotherms for each component and their mixtures.

III. EXPERIMENTAL TECHNIQUES

The main components of the film balance are the trough, barriers, and a device for measuring the surface pressure. While commercial devices are available, many investigators design film balances which embody the principles of either the horizontal float or Wilhelmy plate systems to suit their own experimental purposes. The choice of type of film balance is sometimes a matter of taste but is often dictated by the nature of the interface, as in the case of oil-water [28, 29], oil-air [30, 31], and mercury-air interfaces [32, 33]. Each of these interfaces poses special technical problems in the design of barriers and pressure-sensing devices. However, many technical problems are common to all the measurements in each of these systems. The problems associated with the techniques for studying monolayers at the air-water interface can serve as a useful guide to the solution of similar problems with other interfaces. It can be assumed that the technical difficulties of nonaqueous systems will be at least as complex as those in the air-water system.

A. The Film Balance

1. A Sensing Device for Measuring Very Low Surface Pressures

The two principal methods for measuring surface pressures are the horizontal float system [1-3] and the Wilhelmy plate method [4-6]. The latter method is capable of sensitivities at the 1 mdyn/cm level, but uncertainties arise from the irreproducibility of the wetting of the plate [10]. This is a serious problem when measuring surface pressures less than 0.1 dyn/cm. The horizontal float system does not suffer from the contact angle artifacts as does the plate method, but in the past, optical levers were used which made the measurements inconvenient [3, 34].

A convenient system for routinely obtaining surface pressures with the horizontal float system with a sensitivity of 0.1 mdyn/cm has recently been developed [35]. The unit has two modes of operation, either as a null instrument or as a linear displacement transducer.

The principle of operation of the horizontal float film balance is illustrated in Fig. 5. The float separates two aqueous surfaces, one covered by an insoluble film. As the surface tension of the solution is lowered by the insoluble film, the float moves laterally toward the higher surface tension surface. Operated as a null device, the torsion wire to which the float is attached is twisted to return the float to its original position. The null position of the float is usually noted by the pointer attached to the float assembly shown in Fig. 5. Given the length of the lever arm (i. e., the distance between the pivot point and the water surface), the angular displacement and the mechanical force constant of the wire, the difference in surface tensions, or the surface pressure produced by the film can be calculated readily. In practice, calibration of the torsion wire is accomplished by suspending known weights from the horizontal arm of the float system and measuring the angular displacement.

When used as a linear displacement transducer, the sensor is placed in the position of the null indicator scale (see Fig. 5). The sensor, shown in Fig. 6, consists of an incandescent source of illumination and two semiconductor photodiodes. A vane is attached to the movable null indicator pointer of the horizontal float assembly, and the sensor is positioned so that the vane is between the photodiodes and the incandescent bulb (Fig. 6).

The photodiodes are sensitive only to light which is nearly parallel to their axis and within a cylinder of radius considerably smaller than the radius of the diodes. The vane is slightly wider than the center-to-center distance, so that in midposition it blocks most of the light that would otherwise reach the diodes. A slight movement of the vane then results in a relatively large percentage decrease in the light reaching one diode, and a corresponding increase in the light reaching the other. The current in each diode is proportional to the total light received, so that the current difference due to a displacement of the vane will flow to or from the junction of the two diodes.

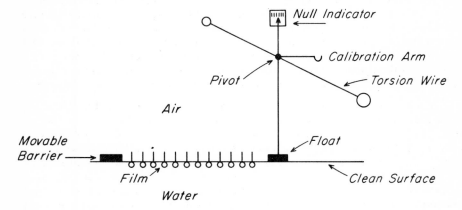

Fig. 5. Diagram illustrating principal components of Langmuir horizontal float film balance

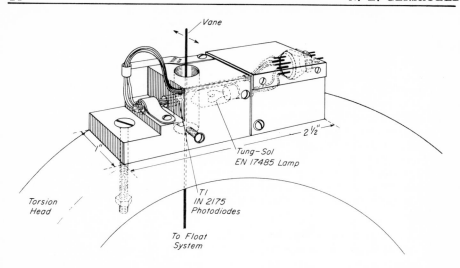

Fig. 6. Scale drawing of sensor fixed in position to torsion head of commercial film balance. Reproduced from Ref. [35] by permission.

This current is converted to a voltage by a circuit using an operational amplifier, as shown in Fig. 7. The feedback resistor can be changed, if desired, to change the overall sensitivity.

Since variations in lamp intensity affect the sensitivity, the power into the lamp is regulated by a 5-V regulated supply and two resistors, so that the source resistance equals the resistance of the lamp at the operating voltage. This makes the power delivered to the lamp independent of small variations in lamp resistance.

For the measurement of very low surface pressures, the displacement of the float is monitored; the displacement will be proportional to the surface pressure. For small lateral displacements of the float, the motion of the vane will also be directly proportional to the surface pressure. Voltage output was found to be linear for displacements of the vane up to 0.3 mm. To stay within the linear range of output voltage for the sensor, torsion wires of different diameters may be used to expand the range of surface pressures.

In operation the device is calibrated by suspending standard weights from the horizontal calibration arm (Fig. 5), and the displacement of the float is read in voltage output. Typical calibration curves for a horizontal float about 10 cm long and a torsion wire diameter of 6 mil are shown in Fig. 8, in which the weights and corresponding equivalent surface pressures in millidynes per centimeter are plotted as functions of the voltage output. Increased sensitivity may be obtained by using a torsion wire of smaller diameter. The absolute error in reading the voltage output is of the order of $\pm 0.1\%$. Under appropriate conditions, surface pressures as low as 1 mdyn/cm can be measured with an accuracy of ± 0.1 mdyn/cm.

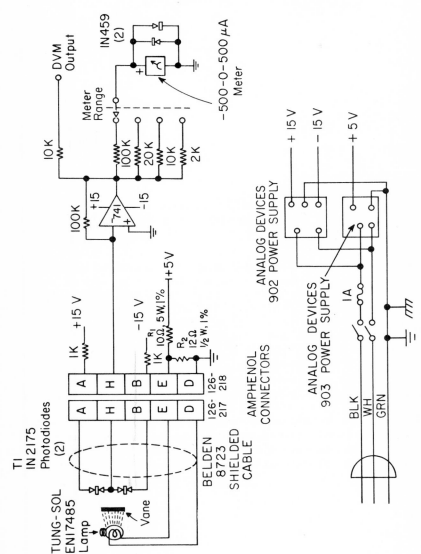

Fig. 7. Schematic wiring diagram of operational amplifier for surface pressure sensor. Reproduced from Ref. [35] by permission.

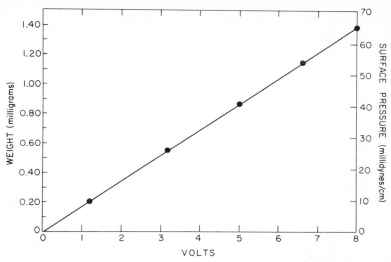

Fig. 8. Calibration curve for sensor. Diameter of torsion wire is 0.006 in.; float length is 10 cm. Reproduced from Ref. [35] by permission.

When the sensor is used as a null device, film desorption or penetration experiments may be performed. In this mode of operation positive or negative displacements of the float from an arbitrary null position may be used to actuate a motor-driven threaded shaft; the shaft, in turn, moves a barrier to either expand or contract the area of the surface to restore the null. The detector sensitivity of film area change will depend on pitch of the thread in the barrier drive shaft, the gearing of the motor to the drive shaft, and the compressibility of the film under study.

2. Float System

The major requirement for measurement of surface pressures in the millidyne per centimeter range is that the float system be essentially frictionless. A float system which has been found useful [36] is comprised of a floating barrier of Teflon strip, 0.0025 cm X 0.5 cm X 10 cm; end loops made of surgical silk which attach the floating Teflon strip to the edges of the trough; and a stirrup which attaches the float to the torsion wire. Most of the difficulties arise from the end loops: If they are improperly attached to the float, they can contribute a significant mechanical drag to the float system. Figure 9 indicates the arrangement which has been found to be essentially frictionless. The end loops are silk threads, 20μ in diameter, which are lightly paraffined. One end of the thread is attached via paraffin to the float, while the other end is attached to a thin Teflon strip which projects out from the edge of the trough. The length of the strip is sufficient to extend beyond the curvature of the meniscus, and its outermost tip is directly behind the extreme edge of the float. The fixed strip itself conforms completely to the configuration of the meniscus. This strip eliminates the possibility

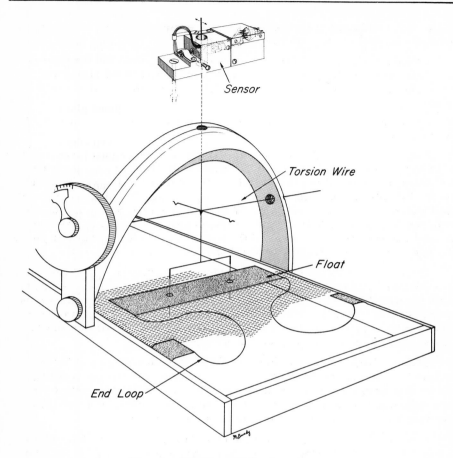

Fig. 9. Diagram of float system

of any leakage occurring at the wall of the trough. The length of the end loop is such that it is allowed to coil naturally behind the float. The actual length of the float from which the surface pressure is calculated is determined by the point at which the end loop is attached to the float (see Fig. 9).

3. Trough and Barriers

The subphase or substrate is contained in the trough, and considerable care must be taken in the choice of materials for its fabrication. The main considerations for choosing the material for the construction of the trough are that it should be nonreactive, should not contribute contamination, and should be easily cleanable.

A variety of materials have been used (e.g., brass, stainless steel, aluminum, polymethyl methacrylate, Teflon, and glass). It has been our experience that all troughs made from metal, even when coated with paraffin or sheets of Teflon, will generate ionic contaminants when water is the subphase. In the case of ionized films, this contamination is a serious problem, particularly at very low film pressures.

Troughs made of solid Teflon pose two problems: (a) Teflon undergoes cold flow and often changes its shape, and (b) it is porous, and the pores often take up solute from the subphase which may be released slowly at a future time.

By far the most useful material is Pyrex glass. One-piece glass troughs can be made easily.

The rim of the trough must be polished flat, and with sharp edges. Rim widths of 5-10 mm are convenient. The surface of the rim must be coated with a hydrophobic material to prevent leakage of water and film over the trough edge. For this purpose, good quality commercial paraffin (mp ~ 60°) can be used without any special treatment. With some practice, a thin coat of paraffin can be applied by heating both the trough and the paraffin to just above the melting point of the paraffin and then carefully painting the paraffin on the surface. The excess can then be wiped off, leaving a thin film on the rim.

The barriers may also be made of glass. They may be coated uniformly with paraffin by heating the glass in an oven above 110°C and then melting the solid paraffin directly on the glass. Alternatively, one can use solid Teflon barriers, provided they are backed by a rigid support. Convenient dimensions for the barriers are 15 cm X 1.0 cm X 1.5 cm. The barrier should extend about 2 cm beyond the edges of the trough.

At least five barriers should be available, two on each side of the float system for sweeping contaminants from the surface, and one for changing the surface area containing the insoluble film. Generally, the surface contaminants which have been removed from the surface are kept behind the sweeping barriers near the extreme ends of the trough.

In principle, if the barrier and the rim of the trough are both perfectly flat and the edge of the rim is sharp, there should be no leakage of the film past the barrier. For most film balance studies this will not be a serious problem. However, when the rate of desorption of slightly soluble compounds is measured, even very small barrier leaks can be serious. To prevent film leaks from this source, two blind holes (diameter <1 mm) can be drilled in the forward lower edge of the barrier, with each hole centered directly over the edge of the trough. When the hole is filled with mercury, the droplet formed will act as a blockage against possible film leaks. To test for film leaks, it is best to use a liquid compound of low surface viscosity (e.g., oleyl alcohol).

4. Temperature Control

The major source of difficulty one usually encounters in regulating the temperature for film balance studies is the cooling effect arising from evaporation of the substrate liquid. In an extreme case, the surface may be as much as 6° cooler than the regulated temperature of the film balance [37].

Harkins [38] was able to maintain temperatures to 0.001° C by circulating fluid through coils in the base of the trough and the walls of the enclosure, while simultaneously maintaining 100% relative humidity within the enclosure to reduce evaporation. However, it is often not necessary to maintain this precise control over temperature. Where room temperatures are stable, it is only sufficient to record the temperature at the surface of the liquid substrate. A surface thermistor should be used for this purpose.

B. Care and Handling of Materials

1. Substrate

The use of a well-defined substrate liquid is extremely crucial. With triple-distilled water (the last two times from quartz) which is collected and stored in a quartz container, one can be reasonably certain that no serious ionic contamination will be introduced. It is generally advisable to avoid using ion exchange resins in the preparation of water because of the possibility that the resin will introduce organic contamination. In the preparation of electrolyte solutions, the major source of contamination will be from heavy metal ions. It is virtually impossible to eliminate all heavy metal trace contaminants without seriously affecting the properties of the electrolyte solution. For example, the use of complexing agents (e. g. , EDTA) can seriously affect the properties of the insoluble film [39]. The use of various types of activated charcoals is also to be avoided since they have been observed to introduce more contamination than they were supposed to have removed [39].

The simplest procedure to follow for the removal of traces of surface-active organic contaminants is to continually sweep the surface of the solution in the trough until no film can be detected when the surface area is reduced to about 1% of its original value. This may take several hours.

For removal of trace heavy metal ion contamination the method of "sweeping" the surface with an ionized insoluble film will remove most of the ionic contamination. At pH 5-8, the alternate spreading and sweeping of four dioctadecyl phosphate films is sufficient to remove the Ca++ from 0.5 liter of 10^{-7} M solution [39]. At pH > 8 behenic acid films will be equally efficient in removing heavy metal ions [40].

2. Deposition of Monolayers

There are two methods for applying a monolayer to the surface. The film-forming material can be added directly and one can depend on the spontaneous spreading of the compound over the surface. However, the precise determination of the surface concentration is difficult, since errors in weighing this small quantity of material can be serious. The second and most common method is to dissolve the material to be spread into a solvent. The solvent should be nonreactive, volatile, insoluble in the substrate, and free of surface-active contamination. Thorough purification of organic solvents entails distillation followed by a single pass over an activated silica gel column to remove surface-active contaminants [41]. A list of the common solvent systems for lipid monolayers is given in Table I. In some instances it is necessary to use solvent mixtures. For example, for alkyl phosphates [42] a mixture of benzene and methyl alcohol (19:1), and for alkyl sulfates [43] a mixture of n-hexane, i-propanol, and water (50:49:1), were found useful.

There has been some controversy about the effects of residual solvent on the properties of the lipid films [44]. It has become evident that much of the difficulty in the use of spreading solvent results from its solubility in the aqueous phase. Thus, the least subphase-soluble solvent will be desirable.

Another possible source of difficulty in the use of the spreading solvent is the fact that many lipids form aggregates in the nonaqueous solvent which dissociate only at very low concentrations [45]. Conceivably, these aggregates may still persist even after spreading. Where possible, it is best to compare several solvent systems whenever the question of a "solvent effect" arises. Occasionally examining the properties of the monolayer as a function of the volume of the solvent will help separate the solvent artifact.

Useful pipets for depositing monolayers have been discussed [46]. Two common types are the micrometer syringe and the self-filling pipet; both units provide equally good reproducibility. However, the calibration of the pipets is essential for absolute data. In practice, calibration should be done with the spreading solvent so that errors of drainage will be accounted for. Calibration must be done by weight for the volatile solvents. Errors in calibration due to evaporation can be minimized by emptying the solvent into a tared beaker containing n-decane; the partial pressure of the volatile solvent is greatly reduced and the calibration of the pipet is easily accomplished.

TABLE I

Common Solvents for Spreading Lipid Monolayers [a]

Solvent	Boiling point, (°C)	Solubility in water, (25°C, g/1000 g H_2O)
n-Hexane	69	0.01
Cyclohexane	81	0.07
Benzene	80	1.8
Chloroform	61	8
Ethyl ether	35	75

[a]Ref. [10].

When depositing the spreading solution on the surface, care must be taken that droplets of solution do not accumulate near the waxed rims of the trough. This precaution is taken to prevent wax on the rim from dissolving and spreading on the water surface. In most instances, the expedient of adding the spreading solution to the surface slowly enough that droplets do not appear will be sufficient to avoid this difficulty.

3. Cleaning of Glassware

Most glassware can be cleaned with a heated mixture of concentrated sulfuric and nitric acids (10 : 1). The use of chromic acid solutions is to be avoided because of the adsorption of chromate ions to glass, which conceivably could be a source of chromate contamination. After acid cleaning, copious rinsing with water is usually sufficient to clean most glassware that has been soaked in acid.

C. Use of Radiotracers for Measuring Excess Surface Concentrations

When applying the Gibbs equation [Eq. (11)], it is essential to obtain values of surface excess concentrations, for Γ. For each insoluble component of the monolayer, Γ is, of course, equal to the amount deposited. However, in the case of monolayer components which are soluble in the subsolution, rigorous evaluation of Γ for the soluble component can be experimentally determined only under special conditions [26].

In the case of a single soluble film component, measurement of the surface tension as a function of the solute concentration is sufficient to determine Γ by applying the Gibbs relation directly. However, for systems which consist of more than one monolayer component, of which at least one is soluble, it is not possible to unambiguously obtain Γ for the soluble component from the Gibbs relation. Perhaps the most general method for evaluating Γ for the soluble component is with radiotracer techniques. The validity of the technique has been discussed [47-49], and a brief outline of the method is given here.

The solute is labeled with a radioisotope which emits weak beta radiation. A detector of the radiation (e. g. , Geiger-Mueller tube for particles with energies exceeding 0.3 MeV, gas flow windowless or thin-window counters for energies $<$ 0.3 MeV) is then placed at a close, fixed distance to the surface of the solution and the radioactivity of the solution R (in counts per minute) is measured. If surface adsorption of the labeled solute occurs, R will be a measure of the radioactivity both from the surface R_S and from the bulk R_B. Thus

$$R_S = R - R_B \tag{15}$$

but

$$\Gamma = R_S / K\rho \tag{16}$$

where K is the detection coefficient (which includes the geometry of the detecting system, counting efficiency, back- and self-scattering factors, etc.) and ρ is the specific activity of the labeled compound (curies per mole).

R_B may be evaluated by measuring the radioactivity R_B' of a solution containing a similarly labeled non-surface-active solute where

$$R_B = R_B' \; \frac{C\rho}{C'\rho'} \tag{17}$$

where C and C' are the solution concentrations of the surface-active and non-surface-active labeled compounds, respectively. K in Eq. (16) may be obtained by measuring the radioactivity of an insoluble monolayer of a similarly labeled compound of known specific activity ρ. Since R_S and Γ are measured, and ρ is known, K is readily obtained.

Thus, three separate measurements are required to obtain Γ. Moreover, it is important to have good values for the relative specific activities of each of the labeled compounds, since errors in their values directly affect the accuracy of Γ. It is therefore recommended that great care be taken to obtain these values directly, rather than routinely accepting the specifications given by the manufacturer.

The sensitivity of the method depends on a variety of factors, some of which may be examined by the easily derived relation:

$$\frac{R_s}{R} = \frac{1}{1 + (C/\Gamma)m} \tag{18}$$

where m is the mass absorption coefficient (cm^2/g) for beta rays in solution. Clearly, the highest accuracies can be expected for small values of C/Γ and for large values of m. C/Γ is a function of the surface activity of the solute, while m depends on the nature of the radioisotope used. Table II lists a group of useful isotopes for monolayer studies, along with approximate values for m.

By far the most useful isotope is ^3H, for which m is about 50 times larger than for ^{14}C or ^{35}S. Until recently, use of tritium as a tracer was limited by the fact that it cannot be detected by ordinary end-window Geiger-Mueller tubes or gas flow detectors. However, Frommer and Miller [50] recently made extremely thin windows (of the order of 10 $\mu g/cm^2$) for gas flow counters, and with this detector they have been able to measure Γ for tritiated DNA solutions.

Another, more elaborate, device for counting tritium at liquid surfaces was developed by Muramatsu and his co-workers [51, 52]; it utilizes essentially a gas flow windowless 2π proportional counter. The detector is a scintillating plastic sheet combined with a prismatic light guide for transmitting the scintillations to phototubes. Humidified hydrocarbon gas must

TABLE II

Radioisotopes for Use in Monolayers

Isotope	Half-life	Radiation	Radiation energy (MeV)	Approx. max. range in water	m (cm^2/g) [a]
^{3}H	12.3 years	β⁻	0.018	6 μ	1.66 X 10^3
^{14}C	5760 years	β⁻	0.155	300 μ	33
^{35}S	87.2 days	β⁻	0.167	340 μ	30
^{36}Cl	3.03 X 10^5 years	β⁻	0.714	~2 mm	5
^{32}P	14.2 days	β⁻	1.71	~1 cm	1
^{45}Ca	165 days	β⁻	0.25	650 μ	15
^{131}I {	8.04 days	β⁻ γ	0.61 0.36	~2 mm	5

[a] Mass absorption coefficient.

be used in this system. The hydrocarbon gas may adsorb to the water surface and possibly alter the surface pressure. This aspect must be carefully checked for each system.

D. Accessory Equipment

Until now, discussion of the film balance has been devoted largely to methods for evaluating the thermodynamic properties of the monolayers. However, it is often useful to examine other properties of the films to supplement the thermodynamic measurements. Some of the other techniques which have been applied to the study of monolayers are monolayer spectroscopy, ellipsometry, the surface potential, and surface viscometry or rheology. The first two methods require elaborate equipment; these techniques will not be discussed, and for details in their use the reader is referred to other sources [53]. Both the surface potentiometer and the surface viscometer are relatively easy to assemble and should be used routinely. An extensive discussion of both techniques has been presented [53]; hence this section will consider only some of the applications which can complement the thermodynamic information.

1. Surface Potentials

The surface potential of the monolayer ΔV refers to the difference between the volta potential of the clean uncovered substrate surface, and the monolayer covered substrate surface. The theory of the surface potential has not been rigorously treated, but a self-consistent approach, which treats the monolayer as an oriented array of dipoles forming a parallel plate condenser, has led to some interesting results [54]. The relation between ΔV and the surface dipole moments may be written as

$$\Delta V = \sum_i 4\pi n_i \mu_i + \psi_o \qquad (19)$$

where μ_i is the vertical component of the dipole moment in the surface and n_i is the number of dipoles per square centimeter of the i th surface component. The presence of ionogenic groups in the surface, when ionized, results in an electrical potential ψ_o at the surface [55].

While some investigators have attempted to interpret ΔV values in terms of group moments, legitimate objections to these conclusions have been raised [56]. The major objection is that most treatments ignore the contribution of the substrate dipoles to ΔV. Since for the present the substrate contribution cannot be evaluated independently, the calculated surface dipole moments can have only marginal theoretical significance.

Despite the limitations, ΔV can be of use for identifying the presence of a phase change in the monolayers [57-59], as well as confirming the similarity of physical states in monolayers for a homologous series of aliphatic compounds [60]. Moreover, dissociation of ionogenic groups in films may be observed, and the surface pK can be estimated [60-63].

2. Surface Viscometry or Rheology

Various types of surface viscometers have been developed [64]. The simplest and perhaps the earliest method for studying the rheological properties of monolayers is to sprinkle particles of talc over the film surface and observe their motion when air is blown gently across the surface [9]. A more sophisticated technique, which is readily assembled, is the torsion pendulum viscometer [65]; it is capable of giving very reproducible results. This technique is useful for identifying phase transitions [66] and reactions in monolayers [67].

The major difficulty in the use of these techniques is in the evaluation of the substrate contribution to the mechanical properties of the film. Though attempts have been made to separate the contribution of the substrate by using canal viscometers, this problem has not been resolved satisfactorily. However, experiments which measure the amount of water dragged along by lipid films indicate that the films may have a significant influence on the mechanical properties of the substrate water [68].

E. Common Artifacts

Apart from artifacts which arise from the use of impure chemicals and the improper maintenance of the film balance, the major difficulty

encountered by most investigators when measuring π-A isotherms is the variability of the pressure readings due to the instability of the film. Monolayer
instability can be traced to two sources:

1. When the insoluble monolayer is compressed beyond the equilibrium
spreading pressure (ESP) of the film, collapse of the monolayer structure
will occur. Above the ESP the monolayers are inherently unstable, reverting
to the excess phase. The reversion process may be slow, in which case it is
often possible to compress the monolayer to pressures far in excess of the
equilibrium value without much apparent film instability. However, the system is not in equilibrium, which introduces uncertainty into the thermodynamic
measurement.

Many investigators have been compelled to exceed the ESP for a variety of
reasons and have also encountered film instability. To correct for this problem the monolayers are usually compressed at a fixed rate. Though the isotherms obtained are often quite reproducible, that does not mean that they are
necessarily correct. This point has been studied [69] with condensed monolayers of stearic acid; it was observed that the isotherm was strongly dependent on the rate of compression. It is best to avoid surface pressures which
exceed the ESP.

2. The second source of film instability arises from the loss of film
material due to desorption either by evaporation, as in the case of some long
chain alcohols, or through dissolution, as is the case for ionogenic systems,
e.g., the fatty acids. One method of correcting for the solubility problem is
to add neutral electrolyte to the subphase in high concentrations to reduce
the solubility of the monolayer. In principle, this is a poor method because
electrolytes will influence the properties of most ionogenic films, and there
is no way of correcting for these chemical effects. A second method which
is more general than the first is to measure the film area as a function of
time at constant π immediately after spreading the film [14, 16]. In static
systems where stirring occurs in the subphase only by convection, desorption is controlled mainly by diffusion. Solution of the diffusion equation leads
to a relation of ln A proportional to $t^{\frac{1}{2}}$. Thus, by extrapolating to zero time,
the area at time of spreading at a particular π may be obtained. Characteristic rates of desorption for some compounds are given in Table III.

IV. GENERAL CONSIDERATIONS

The intent of this chapter was to establish an experimental framework
within which maximum use of the film balance as a tool for measuring
thermodynamic properties of films may be attained. The ultimate aim of
these measurements is to be able to evaluate the intermolecular energies of
the monolayers. For the present, the film balance measurements have one
major limitation — the contribution of the substrate to the structure of the
surface cannot be rigorously analyzed. However, this limitation does not
preclude the measurement of the other energetic contributions to the film
structure; nor does it prevent the use of relatively simple structured films,

TABLE III

Steady State Rates of Monolayer Desorption

Compound	Ref.	π (dyn/cm)	Approximate rate %/min
a. n-Octadecylsulfate	[70]		
pH 2		8	0.05 [a]
3		8	0.2
4		8	1.0
b. n-Octadecylphos-phonate	[70]		
pH 5.8		8	4×10^{-3} [a]
c. Palmitic acid	[70]		
pH 5.8		8	3×10^{-3} [a]
9.2		8	3.0
d. Myristyl alcohol (evaporation)	[23]	35	0.3

[a] Decreasing the length of the hydrocarbon chain, by two $-CH_2$ groups, increases the rate of desorption by approximately a factor of 10.

such as the lipid monolayers, as models of more complex interfaces. For the remainder of this section some useful approaches for evaluating the contribution of intermolecular energies to the structure of lipid films will be considered. Moreover, some preliminary remarks will be made about the applicability of the film balance to the development of physical insights into more complex systems.

The hydrocarbon contribution to the monolayer structure may be obtained by analyzing the behavior of homologous compounds. This approach has been systematically and successfully applied to a variety of bulk systems [71]. When this approach is used, it should be recognized that the functional dependence on hydrocarbon chain length will be a concommitant of the process under study. Huggins [72], for example, has analyzed the bulk vapor-liquid-solid phase transitions of hydrocarbons and has demonstrated that each transition has a unique dependence on the length of the hydrocarbon chain. A summary of some of his conclusions is shown in Table IV. Similar considerations must be applied to any analysis of the phase transitions in monolayers.

One of the major advantages of studying lipid monolayers is that for condensed films the molecular orientation is generally known. This allows

TABLE IV

Dependence[a] of Some Thermodynamic Properties of Hydrocarbons
on the Number of $-CH_2-$ Groups $-n-$

Thermodynamic function [b]	Dependence on $-n-$ [c]
ΔS (f)	$M + Nn$
ΔH (f)	$A + Bn$
ΔS (v)	$X + Yn + Z \ln n$
ΔH (v)	$C + Dn^{2/3}$

[a] Ref. [72].

[b] f = fusion, v = vaporization.

[c] Capitals signify empirical constants.

one to simplify the interpretation of the thermodynamic measurement. For
example, a phase transition between two states when studied as a function of
the hydrocarbon chain length can often indicate the contribution of the hydrocarbon chain to the phase transition, provided that the molecular orienta-
tions do not differ radically from compound to compound. This last point
can be established by surface potential measurements.

For gaseous lipid films, where the molecular orientation and conformation are obscure, the molecular interpretation of the data is correspondingly
more complex than for the condensed monolayers. For example, Langmuir
interpretated the observed linear dependence of the free energy of adsorption
on the number of $-CH_2-$ groups to indicate that the chains are extended and
oriented parallel to the interface [2], but the likelihood that the linear
dependence is the result of several effects led Ward [73] to conclude that the
aliphatic chains assume a coiled conformation in gaseous monolayers.

Bearing these comments in mind, it is possible to evaluate the hydrocarbon contribution to the structure of lipid films. For the analysis of the
polar group contribution, only relative contributions can be ascertained
because of the uncertainty of the substrate contribution; this estimate is obtained by subtracting the hydrocarbon contribution from the total energy [11].

Apart from the general interest in the structure of monomolecular films,
some of the most interesting applications of the film balance have come
from its use for the study of the properties of mixed films (see Sec. II. C).
Unfortunately, many of the results which have been obtained heretofore are
perhaps obscured by the artifacts which have been discussed earlier.

However, it may be anticipated that "two-dimensional" solubility phenomena analogous to those encountered with bulk systems (e. g. , critical phenomena) will be observed. The advantage of knowledge of the orientation and conformation of the film molecules may be usefully employed to elucidate some of the solubility phenomena.

Film miscibility studies also have some relevance to understanding the forces which control the structure of cellular membranes, many of which contain lipids, the mixing of which can be studied with the film balance. In this connection, other membrane-associated phenomena such as cell lysis may be analyzed with the film balance.

In short, the film balance is a highly developed, sensitive tool for studying the thermodynamics of processes in surfaces. Its usefulness is limited only by the imagination of the investigator.

APPENDIX

The Kinetics of Monolayer Desorption

The following is a brief outline of a theoretical analysis of the kinetics of monolayer desorption. It is given in the present short form mainly for the purpose of defining terms which can be evaluated by a computer program, which is also presented in this section. For the complete analysis, one should refer to the original publication [16].

Consider a monolayer of one component which is slightly soluble in the subphase (or slightly volatile). The film slowly desorbs from the surface, and the rate of desorption is measured by following changes in total film area A_T with time at constant π. Figure 3 illustrates the processes involved in the desorption. It is assumed that the rate of desorption is limited by two processes: (a) film dissolution, i. e. , the movement of molecules from the surface into region $\propto$, a very thin region immediately beneath the film; the unidirectional rate constants for dissolution and readsorption are defined as follows:

$$\text{surface} \underset{k_{-\propto}}{\overset{k_\propto}{\rightleftarrows}} \text{region} \propto \qquad (20)$$

(b) diffusion of dissolved film molecules across an unstirred region of thickness ε, into the bulk of the subphase which is stirred only by convection.

Experimentally, it is found that for many desorbing systems at constant π a plot of ln A_T against time is at first variable, but that after 30-60 min d ln A_T/dt becomes constant. If $k = (-d \ln A_T/dt)_\pi$ then at times greater than 30-60 min, $k = k^\infty$ = constant.

The solution of the partial differential equation which describes this system (including both the dissolution and diffusion processes), for the boundary condition that the bulk concentration remains equal to zero throughout the entire experiment, is

$$\frac{k}{k^\infty} = 1 + 2L\,(1+L)\sum_{n=1}^{\infty}\frac{e^{-\beta_n^2\,\tau}}{L+L^2+\beta_n^2} \tag{21}$$

where, by definition,

$$L = k_{-\infty}\,\epsilon/D \tag{22}$$

and

$$\tau = Dt/\epsilon^2 \tag{23}$$

and where β_n are the roots of the equation

$$\beta\cot\beta + L = 0 \tag{24}$$

Experimentally one measures $k = -(d\ln A_T/dt)_\pi$ at each point in the plot of $\ln A_T$ against time, and then divides by the constant value of k^∞ (this usually occurs 30-60 min after the start of the experiment). Plotting k/k^∞ against time, one then compares the experimental curve with the results of Eq. (21), for which L and τ are allowed to vary within reasonable limits.

The FORTRAN program which follows consists of two sections. The first establishes a table for obtaining k/k^∞ as a function of L and τ. The second section converts the experimental data, in the form of A_T and time (in minutes) to k/k^∞ and compares it with the theoretical values of k/k^∞ obtained in the first section, to find L and D/ϵ^2. At the conclusion of the program, examples are presented of the application of these theoretical results for evaluation of k^{eq} (see Sec. II. A. 2). Table V summarizes the notation used in the Fortran program for the experimental parameters.

TABLE V

Fortran Program Notation of Experimental Parameters

Experimental parameter	FORTRAN notation
A_T	AR
$d\ln A_T/dt$	DERIV AR
k/k^∞	DAV
$D/\epsilon^2 = \tau/t$	K
β	BETA
τ	TAU
L	L
t (time, min.)	T

A. FORTRAN Program for Obtaining L and D/ϵ^2

1. Program for Developing Table of k/k^∞ for All Combinations of τ and L

The procedure followed for this section of the program is to evaluate the first 50 roots of Eq. (24), using Newton's method, for each preselected value of L. To obtain each set of 50 roots of Eq. (24), one first estimates the value of the first root, β_1. Values of β_1 for selected values of L are given in Table VI, from which all the initial estimates of β_1 may be interpolated. Values of k/k^∞ are then calculated from Eq. (21) for each L, with τ allowed to vary from 0.001 to 1.0. For this program, values of τ were selected to increase in five equal steps between 0.001 and 0.005, 20 steps between 0.005 and 0.10, and 10 steps between 0.1 and 1.0. Values of L were chosen between 0.1 and 95; the range and increment may be chosen to reflect the physical characteristics of the monolayer [see Eq. (22)] and the resolution of the experimental value for L which one may desire. The corresponding k/k^∞, L, τ values are accumulated in a table and may be stored for use in the second section of the program which follows. Theoretical curves of k/k^∞ plotted against τ for selected values of L are shown in Fig. 10 (see p. 32).

```
C PROGRAM FOR TABLE OF DAV(THEORETICAL) - TAU - L
C DAVTH IS DAV(THEORETICAL)
      IMPLICIT REAL*8 (A-H,O-Z)
      DIMENSION BETA(50),TAU(34),DAVTH(34)
    5 READ 15,KAPPA,XL,BETA1
   15 FORMAT (I5,2F6.3)
      BETA3=0.0
      DO 30 I=1,50
   16 BETA2=(BETA1**2+XL*(DSIN(BETA1))**2)/
    1    (BETA1-DSIN(BETA1)*DCOS(BETA1))
      Y=DABS(BETA2-BETA1)
      IF(Y-0.00001) 17,17,18
   18 BETA1=BETA2
      GO TO 16
   17 Z=DABS(BETA2-BETA3)
      IF(Z-1.5)19,19,20
   19 BETA1=BETA3+4.0
      GO TO 16
   20 IF(Z-5.0)22,21,21
   21 BETA1=BETA3+2.0
      GO TO 16
   22 BETA3=BETA2
   26 BETA1=BETA2+3.1
      BETA(I)=BETA2
   30 CONTINUE
      J=1
   31 TAU(J)=.001
      XL2=XL + XL**2
   32 S=0.0
      DO 40 I=1,50
      TBETA2=TAU(J)*BETA(I)*BETA(I)
      IF(TBETA2-50.0)39,405,405
   39 S=S+DEXP(-TBETA2)/(XL2+BETA(I)*BETA(I))
   40 CONTINUE
```

```
405 DAVTH(J) = 1.0 + 2.0*XL2*S
407 J=J+1
    IF(TAU(J-1)-.005) 45,46,46
 45 TAU(J) = TAU(J-1) + .001
    GO TO 32
 46 IF (TAU(J-1) -.100) 47,48,48
 47 TAU(J)= TAU(J-1) + .005
    GO TO 32
 48 IF (TAU(J-1)-1.0) 49,49,50
 49 TAU(J) = TAU(J-1) + .1
    GO TO 32
 50 DO 51 L=1,31,3
    M= L+2
 51 PUNCH 52, XL, (TAU(N), DAVTH(N), N=L,M)
C ALTERNATIVE: DATA MAY BE STORED AND RECALLED FOR NEXT PART OF PROGRAM
 52 FORMAT (1H ,F6.2,6F10.3)
525 IF(KAPPA) 53, 5, 53
 53 STOP
    END
```

TABLE VI

β_1 for Selected Values [a] of L

L	β_1
0	1.5708
0.1	1.6320
1.0	2.0288
10	2.8628
100	3.14

[a] Ref. 74.

2. Program for Converting A_T and Time t to Obtain L and D/ϵ^2

This section of the program converts the experimental data, in the form of A_T and time t (in minutes), to k/k^∞ and compares it with the theoretical values in the table of the first section of the program to obtain L and D/ϵ^2. A value K_i (=D/ϵ^2) and a value L_j are selected. For this value of K_i a value of τ is calculated for each t of the experiment ($\tau = Kt$). For the value of L_j and the set of τ values calculated, a corresponding set of k/k^∞ is calculated by interpolation of the table in Sec. 1. The difference between these k/k^∞ and the experimental k/k^∞ is squared and summed. This procedure is repeated for all possible combinations of K_i and L_j, where K is allowed to vary from 0.001 to 0.1 and where L varies from 0.1 to 95 as described above. The particular combination of L_j and K_i (=D/ϵ^2) which gives the minimum sum of squares is printed.

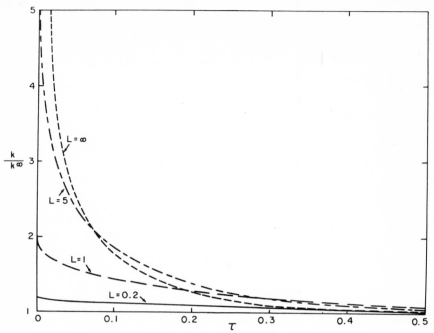

Fig. 10.   Plots of k/k^∞ vs. τ according to Eq. (21).   Reproduced from Ref. [16] by permission.

```
C PROGRAM FOR CONVERTING AR AND T TO EVALUATE L AND K, WHERE
C AR IS FILM AREA, T IS TIME IN MINUTES, AND K IS TAU/T
        DIMENSION AR(50), T(50), ARL(50), DERIV(50), DAV(50),
      1 TABLE(67,33),XL(67),TAU(33),SS(100,67),CARD(20)
        COMMON TABLE,TAU,ISTOP
C THE FOLLOWING 6 LINES PUT INTO THE PROGRAM THE THEORETICAL
C TABLE(DAV(THEORETICAL) - TAU - L) THAT HAS BEEN CALCULATED
        READ (1,890) NL
        DO 760 IL=1,NL
        DO 760 J=1, 33, 3
    760 READ(1,880) XL(IL),TAU(J), TABLE(IL,J), TAU(J+1),TABLE(IL,J+1),
      1 TAU(J+2), TABLE(IL,J+2)
    880 FORMAT (F6.3, 6F10.3)
C IF M=0,PROGRAM TAKES MORE DATA, IF M = ANY NUMBER, PROGRAM OVER
C N IS THE NUMBER OF OBSERVED DATA POINTS
    777 READ (1,890) N, M
    890 FORMAT (2I5)
C CARD IDENTIFIES THE PARTICULAR EXPERIMENT THAT IS TO BE ANALYSED
        READ (1,66) CARD
     66 FORMAT (20A4)
        WRITE(15,67)CARD
     67 FORMAT(' ',20A4)
        READ (1,900) (T(I),AR(I),I=1,N)
    900 FORMAT (2F8.4)
        WRITE(15,749)
    749 FORMAT('0  I        T          AR')
        WRITE(15,748)(I,T(I),AR(I),I=1,N)
```

```
    748 FORMAT(' '15,2F8.4)
        DO 12 I=1,N
     12 ARL(I)= ALOG(AR(I))
C FIND THE SLOPE (DIFFERENTIATE) AT ALL THE POINTS OF ARL
        N1=N-1
        DO 20 I=1,N1
        DERIV(I)=(ARL(I+1)-ARL(I))/(T(I+1)-T(I))
     20 WRITE(15,18)T(I),ARL(I),DERIV(I)
        DERIV(N)=DERIV(N1)
        WRITE(15,18)T(N),ARL(N),DERIV(N)
        N5= N-4
        AV=0
        DO 30 I= N5,N
     30 AV= AV+DERIV(I)
        AV=AV/5.0
        WRITE (15,33) AV
     18 FORMAT(' ',3E15.6)
     33 FORMAT(' AV=',E15.6)
C THE DAV LIST IS THE DERIV LIST DIVIDED BY THE AVERAGE OF
C THE LAST 5 DERIV VALUES
        DO 40 I= 1,N
     40 DAV(I)= DERIV(I)/AV
        WRITE (15,43)(DAV(I), T(I), I=1,N)
        SMALL=99999.0
     43 FORMAT(' DAV=',E15.6,'T=',F8.3)
C THE NEXT SECTION OF CODING,THROUGH STATEMENT 100,FINDS THE
C SMALLEST SUM OF SQUARES OF THE DIFFERENCE BETWEEN
C CORRESPONDING VALUES OF DAV AND THE VALUES FOUND IN THE
C THEORETICAL TABLE,AND THE VALUES OF L AND K FOR THE
C SMALLEST SUM OF SQUARES
        DO 100 IL=1,NL
        DO 90 IK=1,100
        XK=0.001*IK
        ISTART=1
        S=0
        DO 60 J= 1,N
        XTAU=XK*T(J)
        IF(XTAU-1.0) 50,50, 45
     45 Z=1.0
        GO TO 60
C SUBROUTINE INTERP,WHICH IS LISTED BELOW,INTERPOLATES
C BETWEEN VALUES OF TABLE
     50 CALL INTERP(IL, XTAU, ISTART, Z)
        IF(ISTOP.EQ.1) GO TO 885
     60 S=S+(Z-DAV(J))**2
        SS(IK,IL)= S
        IF(S-SMALL)80,90,90
     80 SMALL= S
        LSM=IL
        XKSM=XK
     90 CONTINUE
    100 CONTINUE
        WRITE(15,950) SMALL,XL(LSM),XKSM
    950 FORMAT(' THE SMALLEST SS IS ',E13.6,'WHERE L=' F8.3,',AND K='F8.3)
        WRITE(15,750)
    750 FORMAT('0FOR THE SMALLEST SS,COMPARISON OF DERIVATIVES'/'      J
      1      DAV(J)           Z')
        ISTART=1
        DO 250 J=1,N
        XTAU=XKSM*T(J)
        IF(XTAU-1.0)210,210,200
    200 Z=1.0
        GO TO 250
```

```
210 CALL INTERP(LSM,XTAU,ISTART,Z)
250 WRITE(15,751)J,DAV(J),Z
751 FORMAT(I5,2E15.6)
    WRITE (15,960)
960 FORMAT ('      L      K       SS')
    DO 120 IL=1,NL
    DO 120 IK=1,100
    IF(ABS(SS(IK,IL)-SMALL).GT.0.100*SMALL) GO TO 120
    XK= 0.001*IK
    WRITE (15,970) XL(IL), XK, SS(IK,IL)
970 FORMAT (1H ,2F8.3, E13.6)
120 CONTINUE
885 IF (M) 888,777, 888
888 STOP
    END
    SUBROUTINE INTERP (IL,XTAU, ISTART, Z)
    DIMENSION TABLE(67,33),TAU(33)
    COMMON TABLE,TAU,ISTOP
    ISTOP=0
    DO 11 I= ISTART, 33
    IE=I
    IF (XTAU-TAU(I)) 21,35,11
 11 CONTINUE
    WRITE (15,800) XTAU
800 FORMAT(' ERROR XTAU GREATER THAN ALL TAUS , XTAU=   'E15.6)
    STOP
 21 IF(IE-1)31,31,41
 31 WRITE (15,810) XTAU
810 FORMAT (' ERROR.IE=1. XTAU IS'E 15.6)
    ISTOP=1
    GO TO 50
 35 Z=TABLE(IL,IE)
    GO TO 50
 41 FRAC=(XTAU- TAU(IE-1))/(TAU(IE)-TAU(IE-1))
    ISTART = IE-1
    Z= TABLE(IL, IE-1) + FRAC*(TABLE(IL,IE)-TABLE(IL,IE-1))
 50 RETURN
    END
```

B.   Evaluation of k^{eq} from L

Given the condition that the rate of film desorption is limited by both dissolution and diffusion, it has been shown [16] that

$$\frac{d \ln k^\infty}{d\pi} = \frac{A}{RT} - C_s - \frac{d \ln (1 + 1/L)}{d\pi} \tag{25}$$

but

$$\frac{d \ln k^{eq}}{d\pi} = \frac{A}{RT} - C_s \tag{8}$$

Combining Eqs. (25) and (8) and integrating, the resulting equation yields

$$k^\infty = k^{eq} (1 + 1/L) \tag{26}$$

It is important to note that when $L = \infty$, $k^{\infty} = k^{eq}$. Values of L which exceed 100 are experimentally indistinguishable from $L = \infty$. In Eq. (26), k^{∞} is obtained experimentally, L is computed, and thus k^{eq} may be evaluated.

To illustrate the application of the computer program to actual data, we examine some typical desorption curves. The data for monooctadecyl phosphate monolayers are shown in Fig. 11, where log A_T is plotted as a function of t (in minutes) for $\pi = 8$ and 24 dyn/cm [16]. The data points were treated by the program to search for the best fit with the theoretical k/k^{∞} vs. τ curves (e.g., Fig. 10), and these comparisons are shown in Fig. 12. The best fit of the experimental data yielded $L = 10$ for $\pi = 8$ dyn/cm, and $L = 0.6$ for $\pi = 24$ dyn/cm; $K = D/\varepsilon^2 = 0.01$ for both curves.

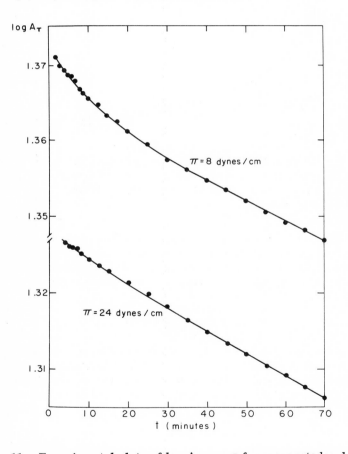

Fig. 11. Experimental plots of log A_T vs. τ for monooctadecyl phosphate monolayers spread at the air–water interface, pH 5.8, T = 22° C. The solid lines were drawn as smooth curves through the experimental points. Reproduced from Ref. [16] by permission.

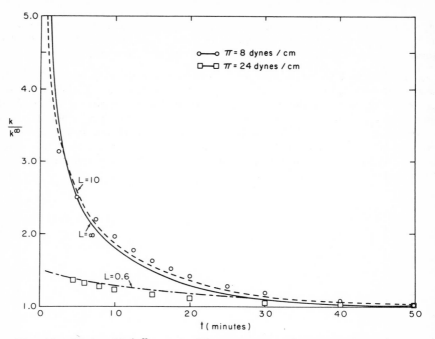

Fig. 12. Plots of k/k^∞ vs. τ. The experimental points are for mono-octadecyl phosphate films spread at the air-water interface, pH 5.8, T = 22° C. The curves were obtained by computer calculation. Reproduced from Ref. [16] by permission.

The values of K $(=D/\varepsilon^2)$ obtained will depend on the hydrodynamic properties of the system. K will therefore be sensitive to temperature control and the uniformity of the temperature in the trough, which greatly affect the intensity of the connection currents in the subphase. It is essential that K be kept constant in all experiments.

ACKNOWLEDGMENT

The assistance of R. G. Minker, M. B. Shapiro, and Dr. C. S. Patlak in the development of this program is gratefully acknowledged. The author also acknowledges Academic Press, Inc. for permission to reproduce Figs. 2, 3, and 10-12, and The American Institute of Physics for permission to reproduce Figs. 6-8.

REFERENCES

1.   A. Pockels, Nature, 43, 437 (1891).

2.   I. Langmuir, J. Am. Chem. Soc. , 39, 1848 (1917).

3.   N. K. Adam and G. Jessop, Proc. Roy. Soc. (London), A110, 423 (1926).

4. L. Wilhelmy, Ann. Physik, 119, 117 (1863).

5. D. G. Dervichian, J. Phys. Radium (7), 6, 221 (1935).

6. W. D. Harkins and T. F. Anderson, J. Am. Chem. Soc., 59, 2189 (1935).

7. Lord Rayleigh, Phil. Mag., 48, 337 (1899).

8. W. B. Hardy, Proc. Roy. Soc. (London), A86, 610 (1912); A88, 303 (1913).

9. H. Devaux, Smithsonian Inst. Ann. Rept., 1913, p. 261.

10. G. L. Gaines, Jr., Insoluble Monolayers at Liquid-Gas Interfaces, Wiley (Interscience), New York, 1966.

11. N. L. Gershfeld, J. Colloid Interface Sci., 32, 167 (1970).

12. L. Ter Minassian-Saraga and I. Prigogine, Mem. Serv. Chim. Etat., 38, 109 (1953).

13. F. M. Fowkes, J. Phys. Chem., 66, 385 (1962).

14. L. Ter Minassian-Saraga, J. Chim. Phys., 52, 181 (1955).

15. N. L. Gershfeld and C. S. Patlak, J. Phys. Chem., 70, 286 (1966).

16. C. S. Patlak and N. L. Gershfeld, J. Colloid Interface Sci., 25, 503 (1967).

17. N. L. Gershfeld, ibid., 28, 240 (1968).

18. J. H. Clint, J. M. Corkill, J. F. Goodman, and J. R. Tate, ibid., 28, 522 (1968).

19. J. Stauff, Z. Physik. Chem., 10, 24 (1957).

20. J. J. Kipling, Adsorption from Solutions of Non-Electrolytes, Academic Press, New York, 1965, p. 211ff.

21. W. W. Mansfield, Australian J. Appl. Sci., 10, 73 (1959).

22. A. Roylance and T. G. Jones, Proc. Intern. Congr. Surface Activity, 3rd, Cologne, 1960, Vol. 2, p. 123.

23. J. H. Brooks and A. E. Alexander, ibid., p. 196.

24. J. H. Brooks and A. E. Alexander, J. Phys. Chem., 66, 1851 (1962).

25. A. E. Alexander and F. C. Goodrich, J. Colloid Sci., 19, 468 (1964).

26a. R. Defay, Thesis, University of Brussels, 1932; see also R. Defay, I. Pregogine, A. Bellemans, and D. H. Everett, Surface Tension and Adsorption, Wiley, New York, 1966, pp. 74-78.

26b. D. J. Crisp, in Surface Chemistry, Suppl. Research, London, 1949, p. 17, 23 .

27.   F. C. Goodrich, Proc. Intern. Congr. Surface Activity, 2nd, London, 1956, Vol. 1, p. 85.

28.   T. G. Jones, B. A. Pethica, and D. A. Walker, J. Colloid Sci., 18, 485 (1963).

29.   L. Blight, C. W. N. Cumper, and V. Kyte, ibid., 20, 393 (1965).

30.   H. W. Fox and W. A. Zisman, Rev. Sci. Instr., 19, 274 (1948).

31.   A. H. Ellison and W. A. Zisman, J. Phys. Chem., 60, 416 (1956).

32.   A. H. Ellison, ibid., 66, 1867 (1962).

33.   T. Smith, J. Colloid Interface Sci., 26, 509 (1968).

34.   J. Guastalla, Compt. Rend., 206, 993 (1938).

35.   N. L. Gershfeld, R. E. Pagano, W. S. Friauf, and J. Fuhrer, Rev. Sci. Instr., 41, 1356 (1970).

36.   R. E. Pagano and N. L. Gershfeld, in preparation.

37.   N. L. Jarvis, C. O. Timmons, and W. A. Zisman, in Retardation of Evaporation by Monolayers (V. K. LaMer, ed.), Academic Press, New York, 1962, pp. 41-58.

38.   W. D. Harkins, in Physical Methods of Organic Chemistry (A. Weissberger, ed.), Vol. 1, Wiley (Interscience), 1945, p. 226.

39.   C. Y. C. Pak and N. L. Gershfeld, J. Colloid Sci., 19, 831 (1964).

40.   J. A. Spink, ibid., 18, 512 (1963).

41.   P. Pomerantz, W. C. Clinton, and W. A. Zisman, J. Colloid Interface Sci., 24, 16 (1967).

42.   N. L. Gershfeld, J. Phys. Chem., 66, 1923 (1962).

43.   J. N. Phillips and E. K. Rideal, Proc. Roy. Soc. (London), A232, 159 (1955).

44.   See Ref. [10], p. 32.

45.   C. P. Smythe, Dielectric Behavior and Structure, McGraw-Hill, New York, 1955, pp. 80-87.

46.   See Ref. [10], p. 62.

47.   D. J. Salley, A. J. Weith, Jr., A. A. Argyle, and J. K. Dixon, Proc. Roy. Soc. (London), A203, 42 (1950).

48.   G. Annionson and O. Lamm, Nature (London), 165, 357 (1950).

49.   G. Annionson, J. Phys. Colloid Chem., 55, 1286 (1951).

50.   M. A. Frommer and I. R. Miller, J. Colloid Interface Sci., 21, 245 (1966).

51. M. Muramatsu, N. Tokunaga, and A. Koyano, Nuclear Instr. Methods, 52, 148 (1967).

52. M. Muramatsu, K. Tajima, and T. Sasaki, Bull. Chem. Soc. Japan, 41, 1279 (1968).

53. See Ref. [10], Chap. 3.

54. J. T. Davies and E. K. Rideal, Interfacial Phenomena, Academic, New York, 1961, pp. 64ff.

55. J. H. Schulman and A. H. Hughes, Proc. Roy. Soc. (London), A138, 430 (1932).

56. A. N. Frumkin and J. W. Williams, Proc. Natl. Acad. Sci. U.S., 15, 400 (1929).

57. N. K. Adam, The Physics and Chemistry of Surface Films, 3rd ed., Oxford Univ. Press, 1941, p. 36.

58. W. D. Harkins, The Physical Chemistry of Surface Films, Rheinhold, New York, 1952, p. 131.

59. N. K. Adam and J. B. Harding, Proc. Roy. Soc. (London), A143, 104 (1933).

60. J. A. Spink, J. Colloid Sci., 18, 512 (1963).

61. J. J. Betts and B. A. Pethica, Trans. Faraday Soc., 52, 1581 (1956).

62. H. C. Parreira and B. A. Pethica, Proc. Intern. Congr. Surface Activity, 2nd, London, 1956, Vol. 1, p. 44.

63. E. D. Goddard and J. A. Ackilli, J. Colloid Sci., 18, 585 (1963).

64. M. Joly, in Recent Progress in Surface Science, Vol. 1, Academic, New York, 1964, pp. 1-50.

65. See Ref. [38], p. 232.

66. See Ref. [59], p. 141ff.

67. N. L. Gershfeld and C. Y. C. Pak, J. Colloid Interface Sci., 23, 215 (1967).

68. C. Y. C. Pak and N. L. Gershfeld, Nature, 214, 888 (1967).

69. W. Rabinovitch, R. F. Robertson, and S. G. Mason, Can. J. Chem., 38, 1881 (1960).

70. N. L. Gershfeld, Advan. Chem. Series, 84, 115 (1968).

71. G. J. Janz, Estimation of Thermodynamic Properties of Organic Compounds, Academic Press, New York, 1958.

72. M. L. Huggins, J. Phys. Chem., 43, 1083 (1939).

73. A. F. H. Ward, Trans. Faraday Soc., 42, 399 (1946).

74. H. S. Carslaw and J. C. Jaeger, Conduction of Heat in Solids, 2nd ed., Oxford Univ. Press, 1959, Table II, p. 492.

Chapter II

MEASUREMENT OF MONOLAYER PERMEABILITY

Martin Blank *

Department of Physiology
Columbia University
College of Physicians and Surgeons
New York, New York

──────
*Supported by Research Career Development Award (K3-GM-8158) of the U. S. Public Health Service.

41

I. INTRODUCTION

The primary aim of this review is to provide a summary and evaluation of the techniques that have been used to measure the permeability of monolayers. The chapter is divided into five sections to allow the reader considerable flexibility in obtaining a sufficiently detailed treatment of any particular technique without having to read through all of the material. Section II presents a historical perspective, not an exhaustive review, that can serve as a reasonable guide to the literature. However, it is not required for the subsequent material. Section III, which deals with the general problems of measurement, is essential for an understanding of the individual techniques. Section IV treats several recommended techniques individually, and only part of this section is usually of interest to any reader. In the concluding section, which is quite short, a number of problems of interpretation specific to monolayer permeability are discussed.

The review is essentially self-contained for the physical chemist having experience with monolayer studies. The less experienced may find it useful to refer to various articles in this collection, but it is not necessary for following the material presented here. Obviously, a review cannot hope to cover all of the details of any particular technique, and the reader will find that the original articles and general references will be of help in correcting these deficiencies. Among the general references Gaines [39] is particularly good for his discussion of monolayer techniques and Davies and Rideal [31] for their chapter on aspects of interfacial transport processes.

Since monolayer permeability is a measure of molecular diffusion through monolayers and is not related to the effects of monolayers on interfacial transport processes that involve convection and stirring, the effects of monolayers on convective processes will be considered only in connection with the problem of measuring molecular diffusion. For a review of the non-diffusion aspects, the reader is referred to Davies [30] or Davies and Rideal [31].

II. HISTORICAL PERSPECTIVE

A. Permeability of Monolayers to Water

Early in the study of insoluble monolayers, it was suggested that compressed monolayers might offer an appreciable barrier to the passage of gases, since they could be compressed to areas occupied by molecules in a crystal. The study that could test this idea most conveniently was the effect of the monolayer on the evaporation of water which formed the subphase. This particular study has increased in importance in recent years, spurred by the desire to reduce evaporative losses from reservoirs [55].

In 1924 Hedestrand [44] attempted to measure the effect of monolayers on evaporation of water into a stream of air. He concluded that monolayers had no effect, but he failed to take into account the stagnant layer of air

immediately above the surface, which was a much greater barrier to evaporation than the monolayer. However, in 1925, Rideal [73] was able to
demonstrate the effect of monomolecular films by using an inverted U-tube,
one side of which contained the monolayer spread on water at room temperature, and the other side cooled in an ice bath. After the system had been
evacuated, the rate of condensation of water in the cold tube was measured
with and without a monolayer. With this apparatus it was possible to overcome the barrier due to the layer of air and to show a sizable effect (28%
decrease) for a stearic acid monolayer on the condensation rate.

In 1927, Langmuir and Langmuir [58] presented a simple way of analyzing the effects of monolayers on the water evaporation rate and introduced
many of the ideas that have come to dominate the field, e. g., the steady-
state analysis in terms of series resistances and the energy barrier theory
for the effect of monolayers. They also showed that monolayers can have
large effects (even greater than by diffusion resistance) in systems where
interfacial transport is affected by stirring near the interface. The particular results in the paper, on the effects of monolayers on the evaporation of
ethyl ether, have pointed out the importance of eliminating nondiffusional
processes near the interface (stirring of liquids and flow of gases) in order to
be able to measure true monolayer permeabilities.

In the late 1930's and early 1940's a number of Russian investigators
(Baranayev and co-workers [4, 82, 83] and Glazov [40] approached the
problem of water evaporation resistance by monolayers with the aid of several different techniques. They studied the evaporation of water and various
solutions under quiescent conditions, when there was a partial vacuum above
the surface or in a stream of air, and they found effects for some surface
films, e. g., cetyl alcohol, but not for others. These workers were influenced by the earlier results of Langmuir and Langmuir [58], and some of
the effects they found for monolayers were ascribed to the ability of the
monolayer to decrease stirring movements near the interface. However,
since they were able to find no effects for films such as oleic acid, a very
permeable monolayer, it is likely that their measurements related to mixed
conditions of diffusion resistance plus the extra resistance of the monolayer
in stirred systems.

Sebba and Briscoe [79] in 1940 attempted to establish the effect of surface pressure on the ability of monolayers to retard evaporation by measuring the amount of water evaporated into a current of dry air. They found
large effects for several films, such as the straight chain fatty acids and
alcohols, but only above the surface pressure at which the film became relatively incompressible. Some films, such as egg albumin, cholesterol, and
oleic and elaidic acids, offered no resistance at all surface pressures. It
is not clear from the paper to what extent the results may have been affected
by the use of an improper spreading solvent or by motion of the monolayer
due to the air current, but their results agree with later data obtained under
unambiguous conditions.

In 1941 Sebba and Rideal [80] reported that films of protein and tanned protein (complexes with tannic acid) have no measurable effect on water evaporation, and that the complex cetyl alcohol-cetyl sulfate film, formed by penetration, has a permeability similar to that of a cetyl alcohol monolayer. They found ethyl stearate less permeable than stearic acid (esters generally being less permeable than the acids) and a mixed film of C_{20} acid and C_{22} alcohol more permeable than films of either component, in line with the idea that the resistances of mixed films on the surface behave like resistances in parallel.

Later on, Sebba and co-workers [81, 47] studied the surface pressure dependence under conditions of partial vacuum, where the evaporation rate was determined by the rate of condensation on a cold tube above the water surface. In this way they were able to overcome the main objections to their previous work, namely, that there was a stagnant air layer above the water surface giving rise to a resistance, and also that the current of air used to absorb the water somehow affected the film. The experiments on cholesterol were repeated and it was checked that even though the film becomes incompressible at very low surface pressures, it has no effect on the evaporation rate. Some work was also done on the resistance of films of docosyltrimethylammonium bromide and docosane-1-sulfonic acid. On comparison with previous results for the same hydrocarbon chain length (a C_{22} chain at a surface pressure of 28 dyn/cm), the order of effectiveness of the end group in reducing evaporation was $OH > COOH > SO_3^- > NM_3^+$, a sequence related to the bulkiness of the polar group, which presumably limits the packing in the film.

In 1943, Langmuir and Schaefer [59] published the results of a new technique, which consisted of suspending a solid desiccant a known distance above the water surface and measuring the uptake of water with and without a film on the surface. The process was analyzed as steady-state diffusion, and from a comparison of the rates with and without films it was possible to determine the evaporation resistance at the surface. Although the technique and analysis were of great value, the actual experimental results were later shown to be in serious error because of improper spreading solvents.

Archer and LaMer [3] improved the technique of Langmuir and Schaefer and determined the specific resistances to evaporation (r) of several straight chain saturated fatty acid monolayers as a function of surface pressure, chain length, monolayer phase, subphase composition, and surface temperature. They also checked the steady-state diffusion equations and demonstrated their validity. They found that in the liquid condensed surface (LC) phase the resistance was independent of surface pressure and subphase pH, and that log r was a linear function of chain length. In the solid surface phase, log r is a linear function of both chain length and surface pressure. It was also shown that log r is a linear function of the reciprocal of the absolute temperature, which is in line with the description of evaporation resistance in terms of an exponential energy barrier. The energy barrier theory was developed and values of the activation energy for evaporation were calculated for fatty acid monolayers and for a CH_2 group in the film.

Rosano and LaMer [74] extended the previous work to include saturated and unsaturated fatty esters and fatty alcohols. Their results showed that the liquid esters (which gave compressible films) had no effect, while the incompressible ethyl stearate was more effective than the stearic acid film. The long chain fatty alcohols were very effective, but for alcohols and esters they found a dependence of r on surface pressure. On the basis of these results they concluded that for monolayers of substances with straight hydrocarbon chains, the resistance to evaporation is determined largely by the surface compressibility. They measured the surface viscosities and found no correlation between resistance and surface viscosity, which ruled out the possibility of surface convection as a factor in these experiments. They also demonstrated that in mixed films, the resistances add in parallel, but not in the case where one component is apparently squeezed out of the film on compression.

During the last years of his life, LaMer and a number of co-workers were interested primarily in the problems of the effects of impurities and the related problem of permeation through mixed monolayers. LaMer and Robbins [57] spread stearic acid films using different concentrations in benzene and showed that the amount of impurity determined from the pressure area characteristics correlated with the resistances reported previously. LaMer et al. [56] and Barnes and LaMer [5] published a number of studies on the behavior of mixed films. These studies added to the available information on water evaporation and were directed primarily to the practical problem of controlling water losses from reservoirs [55].

The amount of information that has been accumulated using the relatively simple Langmuir and Schaefer technique is a good indication of its reliability. There have been few changes in the basic apparatus and only recently has there been any attempt to automate part of the measurement. Ogarev and Trapeznikov [67] have introduced the use of a spring balance to measure the amount of water absorbed continuously, rather than intermittently, but they have retained the basic technique.

Several new techniques have been described recently, but each has serious deficiencies in comparison to the improved Langmuir and Schaefer method. Heller [45] has developed an apparatus to measure water evaporation rates by the temperature changes caused by the condensation of water on metal probes above the surface. A temperature difference is measured between two probes, one over a clean surface and the other over a monolayer-covered one, and this is related to the difference in the rates of evaporation. It is immediately apparent that there would be many more difficulties in working with this system, since two water surfaces must be kept clean (and separated if they are part of the same trough), the metal surfaces must also be kept clean or at least uniform, all air currents and water ripples must be avoided, the distances of the metal probes above the surfaces must be the same and constant during a measurement, etc. It is not surprising, there-fore, that this technique has not been applied widely,

Another technique that offers possibilities for estimating monolayer permeability to water indirectly comes from a study of the stability of small water droplets dispersed in air. Bradley [21] has indicated that water drop-lets of small ($\sim 12 \mu$) radius ought to be very unstable, yet if they are formed with cetyl alcohol dispersed in the water, they last for very long times. Ob-viously, the cetyl alcohol coats the droplet and decreases the evaporation. Recently [35], the evaporation has been studied from droplets ($\sim 100 \mu$ in radius) coated with vapors adsorbed from the gaseous phase (e. g., cetyl alcohol), which offers a better way of controlling the surface film. However, the technique has many serious limitations due to the inability to control the surface film, and the analysis of the process is particularly complicated by the small size of drop, the changing area during a measurement, adsorption and desorption kinetics, etc.

A technique that appears to be particularly well suited to the study of adsorbed, relatively permeable films was described recently by James and Berry [48] and applied to the study of evaporation through protein mono-layers [18]. Thin films of protein solutions were picked up on a ring sus-pended from a balance and the weight was determined as a function of time. Since the water evaporated through the adsorbed protein monolayers, it is possible to use the weight loss rate to estimate the permeability of the mono-layer. This technique requires neither uniform nor complete drainage, and it appears to be insensitive to small amounts of coagulated protein on the surface. However, this technique is limited in accuracy and should only be used when other techniques are not applicable.

B. Permeability of Monolayers to Gases Other than Water

In 1927 Langmuir and Langmuir [58] reported the effect of monolayers on the evaporation of ethyl ether from saturated (5.5%) solutions in water. An insoluble, nonvolatile substance such as oleic acid lowered the rate of evaporation of ether to one-tenth its value with no film present. The effect of the monolayer can be demonstrated by igniting the evaporating ether and then putting a drop of oleic acid on the solution. The oleic acid extinguishes the flames as it spreads to form a monolayer.

Baranayev and co-workers [4, 82, 83] studied the effects of several monolayers on the evaporation of a number of different gases (e. g., HCl, $CHCl_3$, NH_3, H_2S) from aqueous solutions through several monolayers. They found cetyl alcohol to be effective in all cases, but oleic acid was effec-tive for some gases and not for others. Since the evaporation occurred into a moving stream of air, there was bound to be some disturbance of the mono-layer from the gas phase. Surface tension gradients were probably generated in the surface upon evaporation of some of these gases, giving rise to sub-phase stirring in those cases. These factors, which were recognized by the authors, meant that the monolayer resistance was in part due to nondiffusional factors, and the measured resistance could not be related to a permeability.

Sebba and Rideal [80] also obtained results for the effect of monolayers on the evaporation of several solutions (ammonia, ethanol). In the case of ammonia, they reported significant resistances for films of fatty acids, alcohols, and amines, as well as for films of oleic acid, cholesterol, and egg albumin, which have been shown to be ineffective in water permeability studies. It is reasonable to believe that the explanation given above for the effect of an oleic acid film on the evaporation of ethyl ether is valid here. The explanation also applies to the work of Linton and Sutherland [61], who in 1958 reported no effect for a cetyl alcohol film on the dissolution of oxygen under quiescent conditions, but a fairly large effect for the film when the water underneath was stirred.

The first successful attempt to measure the permeability of a monolayer to gases was made by Blank [6], who studied the permeabilities of a variety of monolayers to carbon dioxide and oxygen. By careful examination of the experimental parameters, it was possible to choose conditions in which convection was absent and to demonstrate that the measurements led to an estimate of monolayer permeability. The results of a number of studies [19, 7-9] indicated that the permeability of a monolayer to a gas depended primarily upon the size of the permeant, that there were no solubility or partition effects in the monolayer, and that there appeared to be interference effects between a permeating gas and another nondiffusing gaseous component of the system. The results also indicated that the effects of monolayer chain length and polar group, as well as the effect of temperature, were in line with those that had been observed in the case of water permeation.

The apparatus that was used for these studies was a temperature-compensated differential manometer, and the pressure was measured by the height of a liquid in a U-tube. Recently, Plevan and Quinn [69] developed an apparatus on the same principle using a pressure transducer to allow the data to be recorded automatically. An important point to note in this study is that the authors avoided the problem of convection by using aqueous gels as the gas-absorbing solutions, and their results are in line with those reported earlier by Blank.

Recently, Hawke and co-workers [42, 43] studied the permeability of monolayers to H_2S and CO_2 with the aid of radioactive gases to measure transport. Their techniques has led to values that are substantially in agreement with those reported earlier, and this experimental approach offers considerable promise for the study of processes involving very small amounts of permeant.

A technique that measures monolayer permeability to gases indirectly was reported in 1953 by Brown et al. [23] and was developed and used extensively by Princen and Mason [70]. The technique involves the formation of a soap-stabilized bubble at the gas-solution interface, which after drainage leads to the formation of a film composed of two monolayers and a thin layer of water in between. It is possible to measure the decrease in diameter of the bubble as gas diffuses out and to relate this to monolayer permeability. The method is quite sensitive, but it can only be applied to a limited number of film-forming substances and it requires complete and even drainage of the film. There are also problems in the experiment which complicate the analysis. For example, in order to insure that the gas diffuses through the bilayer into the gas phase and not into the solution phase, the solution is presaturated with the gas by exposure for many hours. However, when the bubble is formed, the gas pressure in the bubble rises (by the Laplace relation) and some gas may dissolve in the aqueous phase at this higher pressure. This apparently does not interfere in the case of the rare gases and oxygen or nitrogen, but it does become a problem for the very soluble gases, carbon dioxide and nitrous oxide. Another problem of analysis is the presence of a water layer of indeterminate thickness between the two monolayers. Despite these problems, the elaborate and elegant work of Princen and Mason has allowed the estimation of monolayer permeability to a number of different gases and has shown that the permeability varies exponentially with the molecular size. While this technique can lead to useful results, it requires great skill and therefore should not be used unless the experimenter is prepared to invest much patient effort.

A more direct method for measuring monolayer permeability, related to the above study, may be possible by an elaboration of a recently published technique. Krieger et al. [53] suspended small gas bubbles in a liquid and measured the size as a function of time to determine diffusion coefficients for the gases in the liquid. If a monolayer can be formed at these gas-liquid interfaces, it may be possible to obtain information about monolayer permeability from these experiments.

C. Permeability of Monolayers to Ions

The problem of monolayer permeability to ions is considerably more complicated because the permeant is influenced by electrical as well as chemical potential differences. Because of the nature of the problem, the monolayers in this case must be present at liquid-liquid interfaces and the liquids must be able to dissolve substantial concentrations of ions. Water is the obvious choice for one of the liquids, but the other liquid presents some problems. One can make aqueous-aqueous interfaces by utilizing the special ability of some polymeric solutions to form coacervates [2], but these systems have such a low interfacial tension (on the order of millidynes per centimeter) that it is difficult to form monolayers there. Another approach has been that of Schulman and Rosano [78], whereby monolayers were formed at the interfaces of a water-pentanol-water system, but because of the large bulk phases and the use of stirring, it was possible to get information only about the ion carrier properties of the surfactants and not about monolayer permeability. An older technique [32], in which a suspended drop of aqueous solution is slowly lowered through a thin layer of a lipid phase that is on top of another aqueous phase, achieves an aqueous-aqueous contact with a bilayer at the interface after the lipid phase is squeezed out. However, this interfacial layer is difficult to control and it is also not very stable. Recently, a number of investigators have been successful in forming stable bilayers between two aqueous phases and in studying their permeability to ions. This work has been reviewed by Tien and Diana [85], and a section of this book deals with the various experimental techniques used in such studies. The analysis of bilayer processes, while useful, is complicated by lack of knowledge of the composition or the concentrations of the various surfactants used to form the layers. For this reason, it is often difficult to draw firm conclusions about monolayer processes.

One monolayer system that has been used to study ion transport is complicated by the presence of an additional material in the interfacial layer [54]. A monolayer is spread at the air-water interface and a thicker layer (~300 Å) of parlodion is layered above it and allowed to set. The resulting film is a monolayer in a relatively thin "inert" matrix that has sufficient strength to support a droplet of water placed on it. The transport of ions between the water droplet and the aqueous subphase can be studied without too much trouble, but the difficulty with this technique is in the interpretation of the results. The actual thickness is not known (there may be swelling and shrinkage) and the monolayer concentration may have changed, etc., so it is difficult to obtain more than qualitative conclusions.

One system that has been used quite extensively in the study of interfacial transport of ions, and which may be of value in the study of the effects of monolayers, is the water-nitrobenzene liquid phase system. The two liquids are immiscible and both can dissolve fairly large concentrations of ions, so that it is possible to study ion transport processes in this system. Davies [29] has studied the partition equilibria of various ions between the two phases and the interfacial transport of ions under concentration gradients in stirred systems. Blank [11, 12, 15] has studied the interfacial concentration changes in ion transport processes across the interface under electric potential gradients, where the charge-carrying species is a surface-active ion. These studies have led to a fair amount of information about the interfacial changes due to current flow across a liquid-liquid interface when one of the ions can form a surface film. It should be possible to form interfacial layers of uncharged substances at this interface and then study the transport of non-surface-active ions due to electric potential gradients across such layers. There are bound to be many technical problems, but it is worth drawing attention to this interfacial system, since a fair amount of information is already available.

The foregoing paragraphs on various methods for determining monolayer permeability to ions indicate the large variety of possible techniques that have been tried or that could be tried. None of the methods discussed has been successful in giving the desired information, but there is always the possibility that someone will be able to modify a method to achieve this aim.

There is one more technique that has been used successfully in a limited number of studies. There are a number of experimental limitations, but the technique is so easy to operate that it may eliminate the need for the development of other techniques. This is the technique developed by Miller and co-workers [38, 63-65] and it involves the use of polarography at a mercury-water interface.

The technique of polarography has been used for some time for the qualitative and quantitative analysis of various substances in aqueous and non-aqueous solutions. The technique is based on the redox reaction of a dissolved substance at the mercury-water interface (at certain levels of polarization), which gives rise to a diffusion current. The qualitative aspect of the reaction is measured by the polarization and the quantitative by the magnitude of the diffusion current. By having a monolayer at the interface in the range where the redox reaction occurs, the effect of the monolayer on the interfacial transport can be determined from the decrease in the diffusion current. This, in general outline, is the experimental approach that has been used to obtain the data about monolayer permeability to ions.

In the case of polyelectrolyte monolayers, the results were analyzed by considering the surface as a separate phase; the transport of the ion through the surface was assumed to be a function of its thickness, the diffusion coefficient of the ion in it, and the distribution coefficient of the ion between it and the bulk phase. In the case of simple surfactants (e. g., decylammonium ions), the analysis was based on a more conventional view of monolayer transport as applied to the dropping mercury electrode. The solution to the problem of an interfacial resistance was given by assuming that the effect of a monolayer on the diffusion current was equivalent in form to the introduction of a rate-limiting step due to a reaction, in this case, the rate constant for crossing the interface [64]. These data allowed an analysis of the effects of different monolayer concentrations and surface pressures on the transport rate and also of the effect of surface potential in controlling the rate.

There are a number of problems of experimental technique and interpretation in connection with the polarographic method [60], but this is the only method which has yielded the desired information about monolayer permeability to ions. The method is useful for uncharged substances, in addition to ions, and can be applied to nonaqueous phases as well. Therefore, in principle, this is probably the most versatile of the techniques described in this article.

III. PROBLEMS OF MEASUREMENT

To study the permeability of monolayers to various substances, it is necessary that the monolayer be present between two phases, L and G (see Fig. 1A) and that a concentration difference of the permeating substance X be established across the monolayer. Furthermore, there must be some way of measuring the concentration of the permeating substance at different times in order to determine the amount of substance transported across the

A. SYSTEM

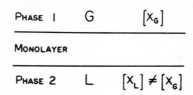

B. MEASUREMENT

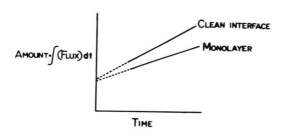

Fig. 1. A. Diagram of a typical transport experiment where a substance X moves from one bulk phase to another across a monolayer. B. The
amount of X transported is plotted as a function of time for a clean interface
and when a monolayer is present.

monolayer (see Fig. 1B). In principle, the experiment ought to be straightforward and relatively simple. One measures the mass transfer, e.g., from
L to G with a clean interface (to determine the properties of L and G) and
when a monolayer is present at the interface. However, there are many
problems that arise in the design of experiments which limit the information
that can be obtained regarding monolayer permeability. For example, the
general difficulties of working with monolayers (e. g., purity of compounds,
spreading, cleanliness of surfaces) become critical considerations because
permeability appears to be especially sensitive to the presence of impurities
that come from all sources, including the spreading solvents themselves.
There are also problems due to the bulk phases, L and G (one of which is
usually aqueous in order to insure the stability of the monolayer), in that
they limit the ability to detect and therefore to measure monolayer properties.
The various general problems associated with the measurement of monolayer
permeability are discussed at some length in this section.

A.   Theoretical Basis of the Measurements

It is obvious that in order to understand the effect of a monolayer it is
first necessary to examine mass transport across a clean interface and

thereby to determine the properties of the bulk phases. The general problem of diffusion of a permeant between two bulk phases, treated by diffusion theory, assumes an unsteady state, molecular diffusion of X between G and L. The solutions to the problem are apt to become quite complicated, especially in systems having special geometries or unusual boundary conditions. If X undergoes a reaction in the liquid phase (the reason for this assumption is discussed in the next section), the problem is further complicated by the diffusion of reactants and products and becomes very difficult to handle. A good summary of the mathematical approaches involved in the solution of diffusion problems can be found in Carslaw and Jaeger [25] or Crank [27].

Assuming that the relevant properties of G and L can be obtained from an analysis of an experiment that has been properly set up to measure diffusion, it is then necessary to repeat the experiment with a monolayer at the interface. If the monolayer experiment is also successful and one can detect a decrease in the mass transfer rate (as in Fig. 1), the data should yield information about the monolayer permeability. But the monolayer cannot be treated as a separate phase for the purposes of the analysis. The influence of the monolayer is included by an adjustment of boundary conditions at the interface. Hawke and Parts [43] define a transmission coefficient at the interface which is decreased when a monolayer is present, and Plevan and Quinn [69] define a mass transfer coefficient through the surface layer which has the same properties. In both cases the adjustment of the boundary conditions is equivalent to the introduction of a resistance (having the cgs units sec/cm) at the interface.

The problem of analyzing the effect of a monolayer becomes much simpler if one can arrange experimental conditions to yield a steady-state system. The steady-state assumption (or approximation) means that the concentrations of X in both L and G near the interface can be taken as constant, and the mass transfer rate across the interface is also constant. In this case the partial second-order differential equations of the unsteady-state diffusion system become simple first-order differential equations, and the solutions are simple algebraic expressions. The effect of the monolayer is generally introduced as an extra resistance at the interface, and the mass transfer is decreased because the driving force (the concentration or pressure difference) must now overcome three resistances (the monolayer and the two bulk phases) in series.

The steady-state analysis, therefore, leads to an expression for the total resistance:

$$r_t = r_L + r_G + r_i \qquad\qquad (1)$$

where the subscripts L, G, and i denote liquid, gas, and interface, respectively [30, 58]. The steady-state analysis is also useful in the study of mass transfer in stirred systems, where it can be assumed that the resistance is concentrated in two stagnant films, one in each phase, of negligible capacity. Because of the stirring, the bulk of the phases has a uniform concentration, the concentration gradients being confined to the two-film region. The resistances of the two films are generally very different in magnitude and usually only one of these films controls the rate.

Let us examine the case of mass transfer between a gas (G) and a liquid (L) phase on the basis of a steady-state analysis. We assume the existence of two films, a gas film and a liquid film (see Fig. 2) and set the mass transport rate proportional to the difference in driving force (concentration or pressure) across each film. The mass transfer coefficient across the liquid film, k_L, is defined by

$$U = k_L (c_i - c) \tag{2}$$

where U is the gas uptake or mass transfer rate per unit area of interface, c_i is the concentration at the interface, and c is the concentration in the bulk liquid phase. Similarly, one can define k_G across the gas film by

$$U = k_G (p - p_i) \tag{3}$$

where p is the pressure in the bulk gas phase and p_i is the pressure at the interface. Since p_i and c_i of Eqs. (2) and (3) are not usually known, one defines overall coefficients K_G and K_L by

$$U = K_G (p - p_e) \tag{4}$$

and

$$U = K_L (c_e - c) \tag{5}$$

where p_e is the pressure in equilibrium with c, and c_e is the concentration in equilibrium with p. The overall coefficients are the ones actually determined from an experiment.

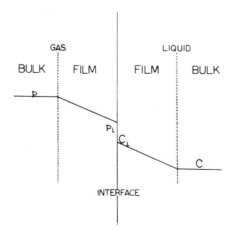

Fig. 2. The steady-state pressure-concentration profile during the transport of a gas across a gas-liquid interface. The symbols are defined in the text.

Since there is an equilibrium between the liquid and gas phases,

$$c = p_e H \qquad (6)$$

where H is the partition coefficient. Combining Eqs. (2) through (6), one derives

$$\frac{1}{K_G} = \frac{1}{k_G} + \frac{1}{Hk_L} \qquad (7)$$

and

$$\frac{1}{K_L} = \frac{1}{k_L} + \frac{H}{k_G} \qquad (8)$$

Since the reciprocals of the mass transfer coefficients can be looked upon as resistances, these equations express the fact that the total resistance to mass transfer (which is the measured resistance) is the sum of the resistances of the liquid and of the gas films, and that the resistances add in series.

If we assume that the resistance of one of the bulk phases is so much smaller as to be negligible, we can express the effect of an interfacial resistance by

$$\frac{1}{K_f} = \frac{1}{K} + \frac{1}{k_i} \qquad (9)$$

where K_f is the total mass transfer coefficient when there is a resistance at the interface, K refers to the mass transfer coefficient (either K_G or K_L) when there is no interfacial resistance, and $1/k_i$ is the interfacial resistance, r_i. We see that Eq. (9) is equivalent to Eq. (1) where either r_L or r_g is negligible. Equation (9), which is valid for steady-state conditions, is also approximately applicable to unsteady-state conditions when the mass transfer coefficients are within a restricted range [6].

If we substitute for K_f and K in Eq. (9) and assume that the mass transfer is from the G to the L phase, with no gas phase resistance ($r_g = 0$), and with an effective zero concentration in the L phase, we obtain

$$\frac{1}{k_i} = pH \left(\frac{1}{U_f} - \frac{1}{U} \right) \qquad (10)$$

where U_f is the uptake for a film-covered surface. This can be rewritten as

$$k_i = U_f / pHx \qquad (11)$$

If we assume

$$x = [1 - (U_f/U)] \qquad (12)$$

Equation (11) was used by Blank [6] in calculating the permeabilities of monolayers to gases for conditions that approximate a steady state. Langmuir and Schaefer [59] used a similar equation for the case of water evaporation studies, under conditions of a true steady state. In the latter case the entire bulk phase resistance was in the gas phase and the driving force was the difference in water vapor concentration across the gas phase (the water concentration above the monolayer minus the water concentration at the desiccant surface).

From Eq. (11) it follows that

$$\frac{dk_i}{k_i} = \frac{dU_f}{U_f} - \frac{dp}{p} - \frac{dx}{x} \qquad (13)$$

This gives the fractional error in k_i on the basis of the experimental variables U_f, p, and x. Since errors in U_f, p, and x may be positive or negative, dk_i/k_i is obtained by squaring each term on the right-hand side of Eq. (13) and taking the square root of the sum. dU/U is the fractional error in the slope of the rate curve and may be taken from Blank's experimental evidence as approximately equal to ± 0.03. dp/p has an estimated value of ± 0.01. dx/x, which generally limits the precision, can be evaluated as a function of U and U_f, from Eq. (12):

$$\frac{dx}{x} = \frac{-(dU_f/U_f) + (dU/U)}{(U/U_f) - 1} \qquad (14)$$

The value of dx/x varies considerably with the value of x. For x = 0.1, dx/x = 0.6, while for x = 0.5, dx/x = 0.1. Most of the experimental data are for the case of x = 0.2, for which dx/x = 0.3. If the last value is put into Eq. (13), dk_i/k_i is calculated to be in the range of 30%, in agreement with the observed value of the standard deviation.

The Princen and Mason [70] measurements were analyzed on the basis of standard non-steady-state diffusion kinetics, since in this case the problem was amenable to such an approach. The physical parameters of the two gas phases were known and the gas phase resistance on either side of the film was assumed to be negligible. The entire resistance was assumed to be due to the three layers in the drained soap bubble — the two surfactant monolayers on the surfaces plus the inner layer of water. On the basis of these assumptions, an expression was derived for the permeability of the entire thin liquid film $\bar{k}$:

$$\bar{k} = \frac{DH}{h + 2D/k_i} \qquad (15)$$

where the water layer of thickness h is permeated by a gas (diffusivity = D, Henry's law solubility in water = H). Although in very thin films ($2D/k_i \gg h$) $\bar{k} = Hk_i/2$ to obtain the monolayer permeability one must generally assume values of h, D, and H. These values are apt to be in error and the calculated k_i is therefore limited in this respect. Princen and Mason calculated their k_i values by the limiting form of the equation, which is also uncertain, and in a more recent publication [71] they have reconsidered some of the difficulties of interpreting the results. One problem that they have not considered is the dissolution of gas from the bubble to the bulk aqueous phase rather than passage through the soap film [18]. This was neglected because the aqueous phase was assumed to be saturated. However, the water is not saturated with respect to the raised pressure inside the bubble once it is formed, and this would tend to accelerate the transport of the more soluble gases (e.g., CO_2, N_2O), as observed. Such a process would also account for the observed changes in k_i with time. This question would require that the results and their interpretation be reconsidered. The original analysis based on Eq. (15) and the variation of bubble size with time appear to be valid, but they ought to be related to the gas that crosses the bilayer side of the bubble only.

The polarographic measurements have been analyzed for a nonsteady system assuming that the monolayer can act by one of two mechanisms. According to both mechanisms the depolarizing ion must diffuse from the bulk of the liquid to the mercury-water interface. However, in the case of polyelectrolytes, the monolayer can be assumed to be a phase with an ion-binding capacity as well as unique diffusion constants. In the case of a straight hydrocarbon chain molecule, this type of analysis is not appropriate and one must rely on a different approach. If one assumes that the monolayer resistance to transport is equivalent to an energy barrier to depolarization at the interface, it is possible to use the formalism for calculating this effect so as to yield the transfer rate constant across the monolayer. This has been done for the permeation of decylammonium monolayers by copper (II) ions [17, 64].

The complete derivation can be found in a paper by Koutecky [52, 22], but for our purposes it is sufficient to calculate the ratio of the diffusion current with a monolayer to the normal diffusion current, i/i_d, which is given a symbol χ. The values of χ are tabulated, and since χ is a function of time t and diffusion coefficient D

$$\chi = k \, (t/D)^{\frac{1}{2}} \qquad\qquad\qquad\qquad (16)$$

one can calculate the rate constant for crossing the monolayer, k.

This approach is relatively straightforward and values are easily calculated from the polarographic data. However, despite the many researches on the polarographic technique, the effects of surfactants are not completely understood [60]. It is for this reason that one must qualify the results of this approach. The effect of the surface film may be due to changes in interfacial flow (affected by a surface viscosity) and changes in the redox reaction rate constants, as well as to the introduction of a diffusion barrier.

B. Detecting Effects Due to Monolayers

In order to obtain information about monolayers from the experiment outlined in Fig. 1, it is essential that the monolayer lower the mass transport rate. This poses a considerable experimental problem because diffusion resistance in bulk phases can easily mask an interfacial resistance. Since diffusion through gases is about 10^4 times faster than through liquids, it is advisable to have one of the bulk phases as a gas and to confine the resistance to liquid phase. (However, a gas can still introduce a resistance in the case of very rapid gas absorption.)

One can completely avoid the problem of diffusion in the liquid by studying water vapor as the gas and having it come through the film from the liquid phase (L) to the vapor phase (G). In this way the water molecules are always present underneath the film. In these experiments, however, the diffusion of the water through a stagnant air layer above the film has often made it very hard to interpret the data in terms of monolayer properties.

In the case of a gas diffusing into a liquid, when the gas has passed through the film it must diffuse away, and this process is quite slow in water at room temperature. For a film to have a measurable effect on the rate at which gas goes into the solution, the film permeability must be several orders of magnitude smaller than that of the water. For the diffusion of carbon dioxide (1 atm pressure) into unstirred water, the uptake

$$U = 2\, pH\, (Dt/\pi)^{\frac{1}{2}} \qquad\qquad (17)$$

For $D = 1.6 \times 10^{-5}\ cm^2/sec$, $t = 100\ sec$, and partition coefficient H taken equal to 1, $U = 0.045\ cm^3$. If the water is covered by a layer $5 \times 10^{-7}\ cm$ thick, for a 25% decrease in rate of carbon dioxide entry ($U = 0.034\ cm^3$), the approximate "operational diffusion coefficient" of gas in film D_f is about $10^{-9}\ cm^2/sec$ or 4×10^{-5} that of water. With increasing values of time, the chances of detecting a 25% effect for the film require still lower values of D_f.

It is possible to use a chemical reaction to remove the gas that has entered the liquid, thereby making room for more gas to come through the film. In the case of carbon dioxide, an alkaline (e.g., glycine-glycinate) buffer reacts very quickly with carbon dioxide ($k' = 7500\ M^{-1}s^{-1}$). Under the conditions given in the above example, if the buffer is 0.1 M, the uptake in 100 sec is 4 cm^3. And for a 25% effect to be observed for the film, D_f need be only 0.004 that of water. Therefore, a rapid reaction for absorbing the diffusing gas considerably improves the chances of detecting an effect due to a film.

Since the effect of a film increases as the uptake rate increases, it appears possible to magnify the effect of a film indefinitely by increasing the rate. However, other effects complicate matters and tend to become important with an increase in uptake rate. The complications are an increase in gas phase resistance during absorption and possible convection effects in the liquid.

C. The Influence of Convection

Explaining the effect of a monolayer in terms of an interfacial diffusion barrier requires that there be no stirring or convection in the bulk phase. If there is convection, the monolayer can minimize the turbulence at the interface, and instead of observing mass transfer due to eddy diffusion, one observes a much slower mass transfer due to molecular diffusion. In general, when there is an inverted density or temperature gradient, there is always the possibility of convection. In mass transport experiments at a gas-water interface, the upper layer of solution can increase in density as a result of gas absorption. There can also be a change in temperature caused by the heats of solution and reaction, which generally tend to reduce the density change and to stabilize the system. In the case of water evaporation studies, there is a fall in surface temperature [49] leading to an increase in density. In any system, therefore, the magnitudes of these effects will determine the stability toward convection.

To give some idea of the magnitude of the effects being discussed, let us consider the absorption of carbon dioxide into a carbonate-bicarbonate buffer. When carbon dioxide is absorbed, the upper layer of liquid forms a saturated solution and the density increases by about 4 parts in 10^4. The change in surface temperature for this case [28] is approximately an increase of $1°C$. This change in temperature should have little effect on the partition coefficient or the diffusion coefficient, and since an increase of $1°C$ in water at $18°C$ corresponds to a density decrease of about 2 parts in 10^4, this reduces the original density increase by about one-half.

In simple systems it is possible to determine the point at which an inverted gradient becomes unstable and causes convection. For water heated from below (an inverted temperature gradient), Rayleigh [72] and Jeffreys [50] calculated the limits of stability, and Schmidt and Saunders [76] have found that the expressions hold quite well. There is a relatively simple expression relating the limit of stability to the onset of convection to the physical properties of the system. For a water layer 0.5 cm thick, the difference in temperature between the top and bottom may be as much as $1°C$; for a 0.1-cm layer it may be as high as $100°C$. Thin layers tend to be quite stable with regard to the onset of convection. If one considers the absorption of CO_2 as in Blank's experiments [6], the absorbing solution should be stable to a difference in temperature of about $1°C$, which, if converted to density, corresponds to 2 parts in 10^4. Although these calculations apply to temperature gradients only, they nevertheless give an estimate of the stability of a system to convection and the magnitude of tolerable density differences.

It is obvious that in order to minimize the possibility of convection a system should be mechanically stable. The temperature should also be controlled to at least $\pm 0.1°C$. (Since monolayer permeability is a highly activated process, temperature control is essential for this reason as well.) Short experimental times also favor diffusion conditions, since it generally takes a while before an inherently unstable system starts to exhibit convection. One very simple way of minimizing convection effects, if this does not unduly perturb the system, is to form an aqueous gel of the phase that is to absorb the gas. This technique was used by Plevan and Quinn [69] and the results indicate that convection was absent. However, Kolthoff and Lingane have indicated that gels do not always eliminate convection [51].

In some systems it is easy to observe convection effects as density changes on the surface (due to changes in index of refraction) which do not vary in a regular way but show eddy-like patterns. After some time, streamers of denser liquid fall through the unreacted solution [1]. In such cases the mass transport data are more erratic and also indicate an increase of rate with time rather than the decrease that is normally expected.

In experiments with monolayers, it is possible to determine indirectly if convection effects are responsible for the observations. In cases where convection effects help to explain the effect of films on the mass transfer rate (e.g., the evaporation of ethyl ether from water), monolayers of oleic acid and cholesterol were found to be effective. In cases where there was no convection (e.g., the evaporation of water), these films had no measurable effect on mass transfer rate. A simple experiment with monolayers of one of these substances should help to indicate the presence or absence of convection effects.

Another type of observation with monolayers may sometimes be helpful in indicating convective effects near a surface that are otherwise undetected. In the case of the absorption of sulfur dioxide by water, the extremely rapid rate of absorption causes visible convective motion at the interface [9]. In this system C_{16} and C_{18} alcohols decrease the mass transfer rate, but the C_{16} alcohol is more effective than the C_{18}. Probably both films are disrupted by the motion at the surface, but the C_{16} is able to spread more quickly to replace collapsed areas and to have a larger effect on the uptake rate. The reversed effectiveness of the fatty alcohols is not observed in systems where convection is absent.

There are a number of ways to show the absence of convective effects when there is no monolayer at the interface. If the expected quantity of mass transported can be calculated, the measured quantities ought to yield values that are reasonably close. If the measured quantities are much greater than the calculated ones, this is probably due to convection effects. If diffusion coefficients can be derived from different experiments, where a substance diffuses into and out of a phase, the same calculated value indicates the absence of convective effects. Even though there undoubtedly were density and temperature changes due to the diffusion of nitrous oxide into or out of various aqueous solutions, the calculated diffusion coefficient was the same in the two directions [68]. Probably the strongest kind of evidence to a surface chemist is an experiment with a very permeable monolayer (e.g., protein, oleic acid). However, it is best to try experiments with a variety of bulk phases to see if the results are consistent. Comparison with other results will sometimes raise suspicions, especially if the monolayers have much lower permeabilities than expected [30].

D. The Need for Mechanical Stability

Monolayers are generally quite stable once formed and will frequently
survive large mechanical disturbances. This is probably due to a variety of
reasons, including: (1) The lower surface tension changes the surface capillary waves; (2) the surface tension gradients produced during a wave (between
points of condensation and rarefaction) cause a diminution in the wave amplitude; and (3) the introduction of a surface viscosity (which is generally considerable) minimizes the movement of the surface under stress. Although the
monolayers are stable enough to be transferred to plates (by the Langmuir-
Blodgett technique), in some cases (e. g. , the S phase of fatty acids, some
high surface concentration protein monolayers) a mechanical disturbance will
frequently disrupt the film. For this reason alone, surface chemists generally attempt to insure a stable mounting and leveling of an apparatus. If
the equipment is to be immersed in a thermostat, and therefore continuously
subject to the flow of the stirred water, it is especially important to insure
stability.

In monolayer permeability measurements, mechanical stability is critical.
One obvious reason is that mechanical instability increases the possibility of
convection, as discussed in the previous section. In most mass transfer
processes, the density and temperature changes at the interface lead to the
onset of convective flow after some time. A mechanical disturbance would
increase the probability of convection and would decrease the time period
before onset. Both of these possibilities are to be avoided, since in either
case the measured effect of a monolayer would lead to an erroneous estimate
of its permeability.

Another reason that mechanical stability is critical in these measurements comes from a consideration of the damping of surface capillary waves
(or ripples) by monolayers. It has been shown [20] that mass transfer rates
across an interface depend upon the amplitude of the ripples and that a monolayer helps to reduce the amplitude of these ripples. Here again, if one is
making measurements in a system where large ripples are present (due to
mechanical disturbance), the measured effect of a monolayer would be due to
a damping of the ripples and would not allow a determination of an accurate
monolayer permeability. Although ripples of very small amplitude are present on a clean quiescent surface, and this poses a theoretical limit to the accuracy of any experiment, they do not constitute a practical limitation in an
otherwise careful procedure.

E. Forming and Maintaining Monolayers

Because of the problems of detecting effects due to monolayers, most of
the experiments have been done with the least permeable monolayers available. These are primarily long, straight chain fatty substances, and it is
generally necessary to dissolve these materials in spreading solvents in order
to form the monolayers. Unfortunately, not all substances are soluble in suitable solvents and the use of solvents that are even slightly soluble in the bulk

phases (e.g., benzene is slightly soluble in water) introduces a serious problem. This problem has been demonstrated experimentally [57] and also analyzed [66], and the conclusion is that a certain concentration of the solvent remains in a mixed monolayer for some time. The presence of the solvent (which is further complicated by being a function of the surface pressure as well as the amount spread and the time) leads to a layer of variable and generally increased permeability. The passage of small molecules across a monolayer may also lead to a decrease in permeability, but this will be discussed in the last section under interference effects.

After considerable work on the problem of solvents, Archer and LaMer [3] suggested a method for spreading monolayers which has proved quite satisfactory. They proposed (a) the use of a very volatile (B.R. ~40°C) water-insoluble solvent such as petroleum ether, (b) a fairly concentrated (~0.01 M) solution of the monolayer substance, (c) the rapid application of the spreading solution, and (d) the spreading of a monolayer at high initial surface pressure. These special precautions are in addition to the normal ones of using specially prepared water (e.g., double-distilled from alkaline permanganate in an all-glass apparatus), avoiding any metals (using only glass that has been cleaned in a chromic acid mixture), and, of course, sweeping the surface immediately prior to the spreading of the monolayer.

It is not always possible to follow these proposals, and, for example, if one is interested in the low surface pressure region, proposal (d) is clearly not relevant. However, they point up many of the problems that must be avoided in attempting to make accurate measurements. There is obviously some leeway, as can be illustrated from the satisfactory use of 0.001 M spreading solutions in 60-80°C petroleum ether, where the solvent was rapidly removed with a vacuum pump after deposition [19].

Archer and LaMer suggested the use of paraffin to coat a metal or glass surface to render it hydrophobic. This method has been used satisfactorily in the past, but there is always the possibility of oxidizing the paraffin if applied with heat, or including solvent molecules if applied in the form of a solution. Both methods might lead to the presence of unknown dissolved material in the subsequently spread monolayer. It should be possible to avoid these problems by utilizing an apparatus constructed of Teflon, which is hydrophobic and therefore does not require paraffin. However, Teflon is soft and the apparatus must be reinforced (externally) to insure stability. Also, commercial Teflon usually contains added plasticizers which are surface active and which must be leached out by soaking for several days prior to use.

To demonstrate the surface effects of dissolved impurities or soluble spreading solvents, one can dust talc powder onto the surface after cleaning or following the addition of solvent. If the surface is clean the talc remains quiescent, but if there are impurities or the solvent has dissolved to some extent during spreading and is now returning to the surface, it is possible to see the characteristic explosion patterns with the talc. (The pattern is similar to the surface wave front produced by throwing a small stone into a pond.) Obviously, this type of observation can be used to check on the cleanliness of the subphase and the suitability of different solvents, spreading times, etc. In all cases where spreading solvents are used, they should be tested for purity by evaporating ~1 ml on a clean water surface and noting the lack of residue. It is also important that mass transport measurements be done with solvent only, to show that they have no effect on the rate (perhaps via cooling the surface upon evaporation).

In discussing the problems associated with spreading solvents, one should not lose sight of the fact that in many cases monolayers can be formed directly from the pure substances. Liquids can spread spontaneously to form monolayers, and some solids (e.g., proteins like serum albumin and even cetyl alcohol to some extent) do as well. Adsorbed films can be formed spontaneously from the solutions to be used, but since these films are generally very permeable, it is extremely difficult to detect any effects due to monolayers.

Ideally, it would be desirable to have some direct measure of the surface pressure or area throughout the mass transport process. Any technique that does not interfere with the transport measurement would be satisfactory, but the Wilhelmy dipping plate technique appears to be the most convenient method and the one most frequently used. The surface pressure generally gives enough information to characterize the monolayer (state, area), and it also serves to insure that the monolayer does not change during the process. In cases where the surface pressure cannot be measured, it is useful to insure that the pressure remain constant by the introduction of a small (bulk) liquid or solid phase of the monolayer substance on the surface. If the bulk and surface phases are in equilibrium, then the surface pressure will remain constant (at constant temperature) and the value can be determined in a separate experiment. In the water evaporation studies this quality of a monolayer has been referred to as "self-healing" ability, and it is particularly important in field applications where a monolayer is subject to conditions that cause the destruction of part or all of the film.

The changes that occur to a monolayer during a mass transfer process should be noted where possible, since they may indicate that the surface has become contaminated. Contamination happens infrequently in a carefully controlled experiment and the more likely observation is the change in the surface pressure due to changes in the temperature and composition of the subphase immediately below the monolayer. The heats of solution and reaction accompanying the mass transport will generally cause temperature changes. The changes may appear negligible in terms of diffusion through bulk phases, since the diffusion coefficient of a gas molecule in water increases by about 2.5%/$^\circ$C. However, in the case of a monolayer, slight temperature increases cause considerable expansion effects and these increase the permeability. In condensed films of C_{19} acid, Archer and LaMer [3] found a change of about 23%/$^\circ$C in the permeability. It is, therefore, important to realize that the measured permeability applies to a temperature range.

Changes in the composition of the interfacial region of the bulk phase can also affect the monolayer. For example, ionization of the polar group is a function of pH, and during the course of carbon dioxide absorption into a carbonate bicarbonate buffer, the layer immediately below the monolayer decreases in pH from 10 to about 8. Fatty alcohols should not be affected, but at pH 10 the COOH group of the C_{18} acid is ionized. Ionization accompanying an increased pH can lead to an expansion of the monolayer, and an increase in permeability has been demonstrated under these conditions [9]. Changes in monolayer packing can also occur as a result of changes in ionic strength.

F. The Polarographic Method

The problems associated with polarography are somewhat different from those of the various other mass transfer techniques described in this article, so it is best to consider them separately. There are a number of very good introductions to the whole area of polarography, e.g., Meites [62], Heyrovsky and Kuta [46], or the somewhat earlier work of Kolthoff and Lingane [51], and they contain much useful information about techniques and general applications. Although the effects of surfactants on the polarographic process have been known for a long time, it is only recently that these properties have been exploited to yield information about the permeability of monolayers. The development and use of this approach has been due to Miller and co-workers [38,63-65] and this information is not yet available in any general review or text.

There are problems that are specific to polarographic methods, such as the cleanliness of the mercury and the capillary tip, the adjustment of the mercury flow rate (i.e., the drop time), the choice of potential range to be scanned, the rate of scanning, temperature control, deoxygenation of solutions, etc. These problems are adequately covered in the general references on the subject, but the problems in connection with monolayers are not mentioned (with the exception of the use of surfactants in the elimination of maxima) and require some comment.

The principle of the polarographic method is quite simple. The mercury-water interface is polarizable and a potential can be impressed across it, with respect to an indifferent (Calomel) electrode in the aqueous phase. If ions are present in the aqueous phase, at some value of the polarization (the decomposition potential) an ion will either give up or receive electrons at the interface (a redox reaction) and therefore undergo a chemical change. Since the ion will have been removed as a result of the reaction, a concentration gradient will be set up and other ions will diffuse toward the interface. In a system with an excess indifferent electrolyte to eliminate electric migration effects, the diffusion of the ions toward the interface can be measured as an electric current. In a polarographic measurement, the mercury-water interface is formed at a dropping mercury electrode where the drop of mercury slowly grows at the end of a fine capillary until it reaches a size at which it drops off. The problem of ion diffusion toward a growing mercury

drop has been worked out, and the (limiting) diffusion current is directly related among other things to the concentration of the ion in solution. For this
reason it is possible to use polarography to determine the concentration of
ions in solutions. A qualitative analysis is also possible, because the potential at which the redox reaction occurs (or the half-wave potential which is
the potential at a value equal to half the limiting diffusion current) is characteristic for an ionic reaction.

To observe the effect of a monolayer, the monolayer-forming substance
is present in the aqueous phase and its surface activity property causes it to
adsorb spontaneously at the interface. There are problems in connection
with this approach since the monolayer may adsorb very slowly in cases of
low concentration or low diffusivity in the aqueous phase as well as unfavor-
able polarizations (close to the potential at which the monolayer is desorbed).
There is the further complication of the growing mercury drop. In the case
of small surfactants (e.g., decylammonium ion), the adsorption can be made
virtually instantaneously, but with large molecules (e.g., proteins) the rate
is slow enough to yield information about adsorption kinetics. For a slow
drop time ($\sim$ 10-15 sec), one can observe the diffusion current at different
stages of the formation of a monolayer from measurements on a single drop.

Another aspect of the problem of insuring that a monolayer is present is
the variation of adsorption with polarization. Since a monolayer is generally
adsorbed at the interface in a limited potential range, it is important that the
diffusion current of the permeant be at its plateau (or limiting) value throughout the range. If measurements are made during the rising phase of the diffusion current, the effects are not easily interpreted. These complications
limit the types of monolayers and depolarizing substances that can be simultaneously studied.

Since the monolayer has to adsorb from the aqueous phase, the solubility
of the surface-active material imposes a severe restriction on the types of
materials that can be used as monolayers. Surface-active ions should be
soluble enough to form monolayers rapidly, but it may not be possible to use
uncharged surfactants of the relatively long hydrocarbon straight chain type.
Hydrophilic polymers have been used in the past and should prove to be no
problem from the point of view of solubility, but they may be denatured or
removed from solution by the deoxygenation procedure. To check on the
presence of a monolayer and to determine its surface concentration as a function of bulk concentration, inert salt concentration, and polarization, it is
possible to make double-layer capacity measurements [17].

We can summarize the above-mentioned requirements for measurements
with this system. Both the monolayer-forming substance and the permeant
must be adequately soluble in the aqueous phase. The monolayer must adsorb
rapidly and in the range of polarization where the limiting diffusion current
can be measured. The permeant, which can be either charged or uncharged,
must have a half-wave potential compatible with the stability of the monolayer.
Since polarography can be performed in nonaqueous media, there is no

reason why the range of possible systems cannot be extended to substances soluble in nonaqueous solvents.

Once the measurements are made, the problems of interpretation become apparent. In many cases, the state (e. g. , solvation, complexing) and the diffusivity of the permeant in the solution are not known, and this information is needed for the calculation of permeability. If the redox reaction is not fully worked out, the ionic species actually involved in the reaction and in the interfacial transport is often not known. Finally, if all these factors are known, the mechanism of analyzing the effect of a monolayer, while apparently satisfactory, still has to be subjected to further testing.

G. Methods of Measuring Mass Transfer

This section contains a brief discussion of the various methods that have been used to obtain the actual data of an experiment. Referring back to Fig. 1 we see that an experiment consists of determining the concentration of X at different times. Usually the choice of X is dictated by the requirements of the experiment, but the method for measuring concentration can vary according to the taste and ingenuity of the investigator. Physical methods have proved much more suitable than chemical methods, probably because they are easier to apply.

An enumeration of the general types of measurement at this point will serve to indicate the broad range of possibilities. In the next section, we shall return to a more detailed consideration of some specific experimental methods.

1. Gravimetric Method

This is the method most frequently used in studying permeability to water. The water is absorbed by a desiccant and weighed periodically or continuously [59, 67]. In a more recent development to deal with adsorbed monolayers, water is allowed to evaporate from a thin film formed on a wire loop (suspended from a balance) and the weight is determined continuously [18].

2. Manometric Methods

The Van Slyke gas analyzer has been used to determine the amount of gas absorbed by a solution in a given time [6]. This experiment gives a single point on the graph of Fig. 1, and although this technique is comparatively accurate, it has been replaced by continuous measurement techniques. The basic simple apparatus is a temperature-compensated differential manometer as designed by Barcroft, with the pressure read on a fluid differential manometer [19]. More recently, pressure transducers have been used to enable automatic recording [69].

3. Optical Methods

In this group we include the simple measurement of water level of condensed transported vapor as reported in the very first successful measure-

ments by Rideal [73], as well as the much more sophisticated "bilayer" bubble diameter measurements to determine gas volume changes [70]. The
broad range of optical techniques (e.g., colorimetry, refractive index
changes determined by Schlieren methods) has hardly been tapped and may
prove to be very useful in special cases [41].

4. Electrical Methods

These measurements can be used whenever ions are being studied and
the ion currents can be measured directly, as in the case of polarography.
The concentration changes can also be determined indirectly by conductance
or potential measurements with ion-specific electrodes. It is interesting to
note that electrical measurements can be used in the study of uncharged
materials such as oxygen [37] by the polarographic technique.

5. Radioactive Tracer Methods

The unique advantage of radioactive measurements is that they can be
used in the study of systems where only trace amounts are involved (e.g., if
a practically insoluble material is to be used as a permeant that must go into
solution). Of course, the technique can be used in all kinds of applications
and has been applied to the study of monolayer permeability to gases [42].

This list is not meant to be exhaustive, and it is important to stress that
a wide variety of techniques can be used. One only has to keep the general
precautions in mind and design an experiment subject to these limitations.

IV. SUGGESTED TECHNIQUES

In this section, several experimental procedures for measuring monolayer permeability are described in detail. Three different groups of techniques are recommended for three different types of permeant — water,
gases, and ions. The procedures that are described have been used successfully and have been found reliable for the measurements. They are, however,
only suggestions, and the individual experimenter may find it necessary to
introduce modifications or even to develop completely different techniques.

While the main points of technique are covered in this section, the general problems of measurement and the necessary precautions are not discussed at any length and the reader should refer to the previous section for
this information.

A. Permeability to Water

1. The Modified Langmuir-Schaefer Technique for Insoluble Monolayers

To use this technique satisfactorily it is necessary to have an adequate
knowledge of the techniques for manipulating insoluble monolayers on a
trough and for determining a compression isotherm. I would recommend the
use of a trough machined from a solid block of Teflon that is screwed (from
the back every 2-3 in. along the periphery) onto a $\frac{1}{4}$-in. metal plate to retain
its shape. The side walls should not be less than about 1 cm thick for

stability, and the inside depth should be great enough to allow the total immersion of a serpentine glass tubing for the circulation of water from a thermostat. Convenient inside dimensions of the trough are 15 X 50 cm. The trough should be mounted on a suitable device that is heavy and shock-resistant for stability, and on three adjustable legs for fixing the level of the surface. The barriers for compressing the monolayer can also be made from Teflon. The bars should be longer than the width of the trough and about 1 cm wide, and their thickness should be sufficient to allow the bar to be anchored by screws onto metal bars.

To measure the surface tension, it is most convenient to use the Wilhelmy method, i.e., a dipping plate attached to an analytical balance [3] , a torsion balance (as in the case of the Rosano tensiometer), or a force transducer. I have found the Sanborn FTA 10 a particularly convenient and accurate transducer that allows the surface tension to be recorded automatically on a strip chart recorder. For a sufficiently sensitive detecting device, a 5-cm-perimeter sand-blasted platinum plate (available through Polytechnics, Oradell, New Jersey) gives a sensitivity of about 0.1 dyn/cm.

The entire apparatus should be assembled in a Vaseline-coated box or hood to minimize the amount of dust that falls on the surface. To measure the isotherm, the surface tension device is calibrated, the surface of the water is cleaned, and the wettable plate is put into the surface. The surface-active material, usually dissolved in petroleum ether, is introduced onto the water surface from a calibrated pipet. The barrier is then moved, either decreasing or increasing the area of the film, and the surface pressure is measured as a function of area either continuously or at regular steps. The surface tension (or surface pressure) plotted vs. the area per molecule gives the compression (or expansion) isotherm of the surface-active material.

To determine the resistance (or permeability) of the monolayer, the evaporation rate of water is measured in the same trough, but with the aid of an additional piece of apparatus (see Fig. 3, which gives a cross-sectional sketch of the components). The desiccant container is a cylindrical Lucite box about 10 cm in diameter. The bottom is open except for a membrane of heavy waterproofed silk cloth (as from umbrellas), which retains the desiccant but is permeable to water vapor. A thermometer, inserted through the rubber stopper, measures the temperature of the desiccant, which is usually anhydrous lithium chloride. An airtight Lucite lid fits the bottom face when the container is not in use. During a measurement, the container fits into a circular hole in the square Lucite platform, which rests on the edges of the surface balance trough. At the edge of the Lucite is a pointed glass rod with which one can adjust the distance between the water surface and the desiccant. The Lucite box and platform are machined so as to hold the bottom face of the container parallel to and a fixed distance (generally about 2 mm) from the water surface. A thermometer is placed just beneath the water surface and another on the floor of the trough.

To determine the rate of absorption, the container filled with desiccant is weighed. The lid is removed and the container put into position above the

A. TOP VIEW

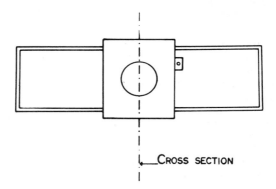

CROSS SECTION

B. DETAILED CROSS SECTION

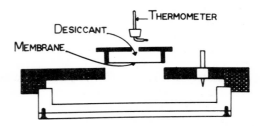

Fig. 3. Diagram of the apparatus used to measure the rate of evaporation of water. A. A top view. B. A detailed cross section along the line shown in A which shows the desiccant container and the platform used to support it. The pointed glass rod, although not along the section line, is also shown in order to illustrate the method of adjusting the height above the water surface.

water surface for a definite time. It is then removed, the lid replaced, and the container reweighed. The increase in mass divided by the area of the membrane and the time in position gives the rate of absorption.

The entire procedure can be outlined in the following steps:

1. The surface is cleaned several times by sweeping.

2. The Lucite platform is put in place and the water level adjusted with the aid of the hydrophilic point.

3. The platform is removed and the film is spread on a freshly formed surface (made by pushing a movable barrier away from a stationary barrier that it originally touched) at a high surface pressure.

4. The Lucite platform and the dipping plate are replaced, and the surface pressure is adjusted.

5. The temperature of the desiccant is measured and the box is weighed.

6. The box is opened and placed on the platform and simultaneously a stopwatch is started.

7. After a time (usually several minutes), the box is removed, closed, and weighed and the temperature is again noted.

From two determinations of the rate of absorption, one with a monolayer on the surface and one without, the resistance of the monolayer can be calculated from Eq. (18),

$$r = (w - w_0) \left[\frac{1}{U_f} - \frac{1}{U} \right] \tag{18}$$

which is a form of Eq. (10). In Eq. (18), w is the concentration of water vapor in equilibrium with the liquid water, and w_0 is the concentration of water vapor in equilibrium with the desiccant. In the system under discussion, evaporation takes place at the interface, and the water vapor molecules diffuse from a concentration w, through a stagnant air layer, to the desiccant where the concentration is w_0. In a steady-state system the rate of evaporation is equal to the rate of absorption and is equal to Eq. (18), which refers to the rate of diffusion across the stagnant air layer.

While both U and U_f can be measured with precision, the values of w and w_0 are known as functions of temperature and must be taken from tables. To insure that proper values are used, the temperature must be accurately known and corrections must be made where possible. There are temperature changes during an experiment, since the water surface cools, the desiccant temperature increases, and the temperature gradient in the air layer varies as a result of these changes. Corrections can be made, and to do this it is assumed that the average temperature at the water and desiccant surfaces during an experiment can be used to evaluate w and w_0.

To determine useful average temperatures it is necessary to make several measurements and also to keep conditions as stable as possible. In order to minimize temperature variations in the water surface, water from a thermostat is pumped through a Pyrex tube inside the trough. The temperature of the surface is determined from an estimation of the temperature gradient in the water. A thermometer on the bottom of the trough measures a nearly constant temperature T_c. The temperature T_t, measured by another thermometer beneath the surface, is assumed to be the mean temperature of the gradient between the plane at temperature T_c and the surface. The temperature of the surface is, therefore, given by

$$T_s = 2T_t - T_c \tag{19}$$

Since the total variation is generally small, Eq. (19) gives a reasonable value for T_s within about $\pm 0.1^\circ C$. This value is then used to obtain the value of w to be used in Eq. (18).

The average temperature at the desiccant surface is also evaluated by Eq. (19), only in this case the temperature is more variable during a run. Therefore, readings must be taken of T_t (the temperature read after shaking the contents of the desiccant) and T_c (the temperature of the upper layer of desiccant normally read by the thermometer) immediately before and after the run. The average values of T_t and T_c are then used to evaluate T_s. This method leads to an uncertainty of less than $\pm 1^\circ C$, which causes a small enough effect on the value of w_0 and is within the precision of the method.

Another correction that must be made before the calculations is the subtraction of the amount of water absorbed by the desiccant from the air while the desiccant container is being put in place above the surface and is being removed. This correction can be obtained by doing a run under standard conditions with the water surface covered by a thin aluminum foil. The magnitude of the correction is generally small, but a careful standardization of the method has shown that it is a necessary correction.

In order for this method to give reasonable values for the resistances, it is absolutely essential that diffusion conditions prevail and that there be no convective motion in the air layer between the water and the desiccant. Archer and LaMer [3] have shown that the proper conditions are present, and they have also tested the experimental limits of the method. For example, they have determined how thick an air layer can be used before convective effects occur. The temperature changes that occur in the air column apparently help to stabilize the system, and in general, the system seems to be quite stable. Because of the relative stability of the system, it may be possible to use a spring balance to suspend the desiccant [67], but this raises a number of questions about this system that ought to be carefully looked into before adopting this procedure. For example, the distance between the desiccant and water surfaces is critical in the measurements and the use of a spring balance necessitates constant readjustment of this distance. This procedure minimizes the weighing problems but adds to the level adjustment difficulties. Another aspect of the level adjustment that raises problems is that the procedure may generate vibrations and convection currents in the air layer that may upset the steady-state diffusion conditions essential for the analysis of the data. For these reasons it is suggested that such innovations be carefully checked for each set of experimental conditions (e.g., ambient temperature, height of air column) to insure that proper conditions are maintained.

2. The Thin Film Method for Soluble Adsorbed Substances

The transport of small molecules across relatively impermeable long chain fatty acids, alcohols, etc., has been studied by techniques that are sufficiently sensitive to enable measurements under a variety of conditions. For adsorbed monolayers, which have a considerable amount of open space, the normal techniques are not sensitive enough. The Princen and Mason

technique for investigating relatively permeable monolayers of soluble substances, which involves the formation of a soap-stabilized bubble at the solution surface, is quite sensitive, but unfortunately it cannot be applied to films
where drainage is uneven or incomplete or if there is a considerable amount
of coagulated material in the film, as in the case of proteins. The technique
that appears to be best suited for the study of such films was described recently by James and Berry [48]. Thin films of protein solutions were picked
up on a ring and suspended from a balance, and the weight was determined as
a function of time. Since the water evaporated through the adsorbed protein
monolayers, it is possible to use the weight loss rate to estimate the permeability of the monolayer [18].

The apparatus required for this technique is quite simple. A fine (0.042-
cm-diameter) phosphor bronze wire or other hydrophilic material can be
made into a 1-cm-diameter ring and suspended from a microbalance or force
transducer (see Fig. 4). Actually, the electric microbalance, EMB-1, of
Research and Industrial Instruments Co., has been used satisfactorily where
the output was fed into a recorder in such a way that full-scale (7 in.) cor-
responded approximately to 100 μg in this system. Any arrangement giving
an equivalent precision should be adequate.

The ring, while on the balance arm, is dipped into a small volume (about
20 cm^3) of the film-forming solution and the solution container is removed.
The balance potentiometer is adjusted and the weight of the ring and solution
is recorded as a function of time. A series of about eight films is formed;
the first two are not recorded but are used to adjust the conditions in the balance case to steady value. The recordings run for at least 1 and generally

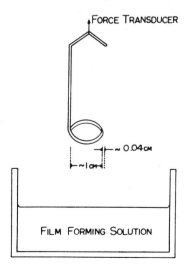

Fig. 4. A diagram of the ring (and its suspension) used to form thin
films of monolayer-covered solutions.

over 2 min and the steady-state slopes at 1 min time are measured. The slopes are averaged and the mean is generally good to about $\pm 3\%$.

The ring and solution container are then rinsed with triple-distilled water at least 10 times, and water at the same temperature is left in the container. The same measurements are then repeated as soon as possible on thin films formed from this solution. The traces of surfactant left by the rinsing are usually enough to insure the stability of the thin films, but their concentrations are such that the evaporation rates recorded could be ascribed to "pure" water. This is the simplest way to determine the evaporation rate from a "clean" surface under conditions that are similar to the experiments with surfactant films.

The resistance of the film is calculated using the steady-state Eq. (18) given earlier, only in this case Δw is the difference between the water concentration at the interface and the water concentration at the bulk of air (just beyond the stagnant layer). Therefore, to determine r, we require two measurements of the steady-state rate and we must know the two values of the water concentration to give Δw.

The steady-state rates can be determined from the known dimensions of the ring and the weight/time measurements, but the values needed to calculate Δw must be estimated. The concentration at the interface varies with the temperature of the solution, and an initial value is known. However, the temperature must fall as the evaporation continues, and for more accurate measurements a method must be developed to estimate this change and to compute an average temperature. This can probably be accomplished by determining the temperature of a relatively small volume of solution that forms the film before a run and again when the ring is dipped in after a run is complete. One of the reasons for using at least 20 cm^3 of solution is to insure that the change in initial temperature is negligible in a series of runs.

The water concentration in the bulk air phase is determined by the humidity (in the local environment) and this can also be determined. However, the humidity changes during a run and it is more important to stabilize the humidity in order to yield reproducible values for the measured rates. This can best be accomplished by doing the measurements in a box or balance case and using a series of runs as described above. By repeating the "clean" surface measurements immediately after the monolayer-covered surface runs, one also minimizes variations in humidity.

Other experimental factors should be controlled in order to insure standard conditions and to minimize the possibility of convection. The ring should not move about once the film is formed, and it should be horizontal. Also, the ring should be rigid so that its area is constant, and it should be made of fine wire so that there is very little change in area as the film drains. The last two requirements are somewhat conflicting, but the phosphor bronze wire is a reasonable compromise. (In connection with the problem of film drainage, it should be noted that the shape and area of the water layer must be corrected when the technique is refined.)

This experimental system has not been used extensively, since it is relatively new, but it is worth developing because of its simplicity and because it appears to make possible the study of very permeable soluble films. In the experiments that have been reported, the assumptions made in evaluating Δw were such that maximal values of the resistances could be estimated [18]. This information can be useful under certain circumstances. If a better evaluation of the Δw is made by better estimation and control of the temperature of the solution and the humidity of the surrounding air, the estimates ought to improve. The actual data for the experiments appear to be in line with the results of Princen and Mason [70] when better estimates of these quantities are used.

B. Permeability to Gases

1. The Modified Temperature-Compensated Differential Manometer

The apparatus is based on the Barcroft temperature-compensated differential manometric apparatus, the detailed theory of which is discussed by Dixon [36]. The modified apparatus, shown in Fig. 5, consists of a gas absorption cell and a dummy cell (of the same size and shape) connected by

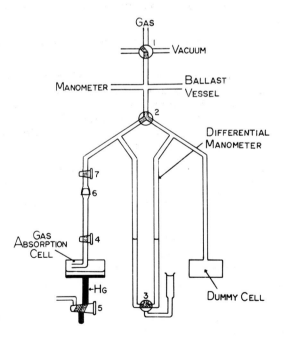

Fig. 5. A diagram of the apparatus used to measure monolayer permeability to gases. The functions of the various parts of the apparatus and the detailed manipulation of the taps are explained in the text.

a differential manometer, as well as taps used to fill the cell, evacuate the system, and control the composition and pressure of the gas phase. The cross-sectional area of the cell should be made large enough to give a rapid absorption rate and yet not make the cell too bulky ($\sim$35 cm^2).

An analysis of the apparatus and a derivation of the formula (following Dixon) for the conditions of these experiments leads to a relation between u, the volume of gas absorbed, and h, the height of the liquid in the differential manometer. The actual relation depends upon the volumes of the various parts of the apparatus and the density of the manometric fluid. (Distilled water that had been degassed was used as the fluid, since it leads to relatively large changes in height and, therefore, a greater precision. A trace of detergent was added to insure proper wetting of the capillary and effective drainage.)

The two cells and the ballast vessel are mounted in a thermostat with the differential manometer mounted in front of the thermostat for ease of reading. Although all of the apparatus should be in the thermostat, it is most convenient to keep some parts out, such as tap 4, joint 6, and the differential manometer. In any case this involves much less than 1% of the total volume. While it is taken for granted that the apparatus must be stable, particular care should be taken with the cell to minimize the effects of the vibrations from the thermostat.

The procedure in each experiment may be summarized in the following steps:

1. Taps 4 and 5 and ground-glass joint 6 are lubricated and set into circuit. (The usual tap greases cannot be used, since they are soluble in the spreading solutions in which the monolayer substances are introduced, and thus might contaminate the films of the latter. Glycerine, which is both water-soluble and non-surface-active, is a satisfactory alternative. It may also be possible to use Teflon taps and joints.)

2. With tap 1 connecting the vacuum pump to the system, taps 2, 4, and 7 open, and taps 3 and 5 shut, the entire system is evacuated.

3. Taps 4 and 7 are closed and the cell is disconnected from the system at joint 6.

4. The cell is filled, via a mercury burette, with a known volume of solution (about 20.0 ml) followed by a suitable known volume of mercury through tap 5. The mercury is added to insure the quantitative transfer of solution and also to keep the cell from floating in the thermostat after evacuation.

5. The cell is reconnected to the system at 6 and the system is evacuated (degassing the solution) with tap 7 open.

6. Tap 2 is closed (all three junctions disconnected) and tap 1 adjusted to let gas come into the ballast vessel. Tap 1 is then turned to connect the vacuum pump to the gas feed, thereby isolating the system.

7. At least 2 min are allowed for the gas to reach equilibrium, and then the pressure is adjusted by bleeding the excess through tap 1.

8. The experiment begins when tap 2 is opened, permitting the gas to flow into the absorption and dummy cells, and after 6 sec tap 2 is shut. (By 6 sec the pressure has settled down to a steady level.) The final pressures can be arranged to be in the range 0.06–0.8 atm.

9. Tap 3 is then opened (as in the diagram) and the height of the column in the differential manometer is recorded as a function of time.

When films are used the following modifications are made after step 5:

a. Taps 4 and 7 are closed and the cell is disconnected from the system at tap 6.

b. About 0.5 ml of spreading solution (usually about 10 mg/100 ml of solvent) are added to the cell and leaked in through tap 4, which is then closed. When the spreading solution hits the surface of the solution, a surface "explosion" characteristic of spreading should be seen.

c. The cell is reconnected to the system at 6 and the system is evacuated with tap 7 and then tap 4 open, thus pumping the solvent away and leaving the film plus a small solid bulk phase, due to the amount of added solution being in excess of that required to produce a saturated monolayer at the interface.

The experiment is then done, commencing with step 6, in the same manner as described above.

After step 9 in the procedure, the measurements of the height of the column in the differential manometer at various times are converted to the volume u of gas absorbed as a function of time. The slope of u(t), which is signified as U, is determined as the linear portion in the vicinity of 1 min. (The departure from linearity does not occur, as a rule, until after this region.) The permeability is calculated from these data and a knowledge of the area of the surface, the pressure of the gas, and the known solubility of the gas in the aqueous phase. The formula used in Eq. (10), which assumes a steady state, no gas phase resistance, and diffusion of the gas from a pressure p to a pressure of zero in the bulk of the liquid. When these assumptions cannot be made, the unsteady state treatment of the data [69] is necessary.

The experimenter should have some familiarity with the techniques of degassing solutions and of manipulating them so as to eliminate the possibility of contact with the air. The simplest device for filling the cell [step 4 of the above procedure] is shown in Fig. 6; it has a bottle containing the solution (I), a mercury burette (II), and the cell (III). The bottle has a stopper containing two glass tubes connected to flexible tubing that can be clamped. When the bottle is sealed it can be evacuated through one tube (degassing accompanied by vigorous agitation of the solution in the bottle) and flushed through with an inert gas (such as nitrogen) via the other tube. After this

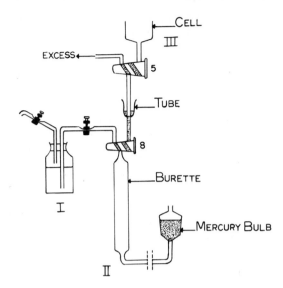

Fig. 6. A diagram illustrating the method by which the absorption cell is filled with solutions that must not come in contact with air. The functions of the various parts of the apparatus and the detailed manipulation of the taps are explained in the text.

process is repeated several times, the solution can be transferred to the cell via the burette.

To accomplish the transfer, the flexible tube (connected to the glass rod that dips into the solution) is connected to the side arm of the burette. The burette is clamped in a vertical position, tap 8 is lubricated with glycerine, and some mercury is moved into the tube above tap 8 by raising the mercury bulb. The tap is closed. The glass rod on the bottom of the cell has a perforated rubber tip over it, and this is pressed into the bottom of the tube, making a seal. The cell is then clamped into place with tap 5 open to the sidearm. The clamp on the bottle is opened, tap 8 is opened to the sidearm, and the mercury bulb is lowered. This draws solution into the burette. Tap 8 is then opened to the tube connection so that when the mercury bulb is raised, the solution can be raised up through the sidearm of tap 5. This procedure is repeated several times in order to rinse the various pathways and to fill the tubes up to tap 5 with the solution. To make a quantitative addition of solution to the cell, the mercury-water meniscus is read in the burette, tap 5 is slowly opened to the evacuated cell, and the mercury bulb is slowly raised until the proper volume of solution has entered the cell. Tap 5 is then opened to the sidearm and the mercury is raised up to the sidearm. At this point a known volume of mercury can be added to the cell, by filling to a predetermined level in the cell or to a predetermined height of mercury as measured by holding the leveling bulb up to a fixed point on the burette. The result of this procedure is that a known volume of the solution has been transferred to the cell and the cell can now be set into the apparatus and the gas absorption rate measured (with or without a monolayer). The added solution and mercury increase the weight of the cell and help to stabilize it, while the mercury in the tap helps to maintain a proper seal.

The cell used in these measurements has an attached right angle tube inside that connects the upper tap (4) to the central compartment (see Fig. 5). This tube conveys the monolayer-forming solution into the cell and allows the addition to come at one point of the surface only. The addition of solution can be better controlled because there is less loss of solution than would occur if the solution were allowed to go down the sides of the cell. There is also the distinct advantage that the monolayer solution is present in a single lens and that a single solid phase is formed when the solvent is evaporated off. This allows a reasonable control of spreading volume and size of excess phase, which are important. The size of the solid excess on the surface must be kept very small so as not to interfere with the mass transport measurements and, in all cases, it must be shown that the measurements are independent of the size of the excess phase in the range of experiments.

The tube in the absorption cell serves another important function at the start of an experiment. When tap 2 is opened, the gas that rushes into the cell can disturb the surface considerably, but the tube minimizes this by acting as a baffle. Even with the tube there is bound to be some effect of the inrushing gas on the surface, unless particular care is taken in the opening of tap 2. The experimenter can generally control tap 2 well enough after some practice, but part of the problem can be eliminated by designing the apparatus with a relatively large bore tap (2) and a relatively small bore tubing connecting the cells and manometer, and by maintaining almost equal volumes and pathlengths for the two halves of the system.

In many experiments, it is necessary to take readings at very short intervals and it is difficult to read two instruments, a clock and a manometer, with any accuracy under these conditions. A useful and inexpensive piece of equipment in this situation is a metronome, which can be calibrated to click every few seconds and which therefore frees the eyes to concentrate on the manometer. The problem is not completely solved, however, because the data must be recorded. But it is possible to write the readings down without looking. I have found it relatively easy to take readings every 6 sec by this method.

This last paragraph brings to mind the question of automating the measurements. Plevan and Quinn [69] have built an apparatus that incorporates a pressure transducer in place of the differential manometer, and they have been able to make measurements every 2 sec. However, this does not improve the accuracy because the limiting factor is still the time required for the gas to be distributed throughout the apparatus, and the precision appears to be comparable to the results reported by the earlier technique. The technique described in this chapter has the advantage of using apparatus that is much cheaper and easier to assemble and that also probably presents fewer problems during operation because of its simplicity.

Nevertheless, for those about to start permeability measurements, it is advisable to study the work of Plevan and Quinn, since their paper contains many useful ideas.

2. The Thin Film Method

The thin film method, first used by Brown et al. [23] and developed by Princen and Mason [70], represents a useful new technique for obtaining information about monolayer permeability. However, the method is indirect, since the measurements are made on layers that are composed of two mono-layers plus a thin layer of water between them. This means that the information obtained from these measurements must be extrapolated to the case of monolayers. This may be a valid procedure, but the presence of the water layer of unknown thickness and the possible differences between a monolayer on a bulk phase and a monolayer in a "bilayer" make the problem of extrapolation difficult.

There are also experimental problems with this method, such as the definition of surface tension and the determination of the area occupied by molecules at the surface. The problem of solubility of the gases in the bulk aqueous phase and the possible convection effects because of the shape of the gas-aqueous interface raise questions about the interpretation of the data. For these reasons and also because the technique is quite difficult to use, the thin film technique is not recommended for making permeability measurements on monolayers, and it will not be discussed further.

Nevertheless, one cannot help but be impressed with the skill of the experimenters in overcoming many problems and in presenting many useful and important data for those interested in problems of interfacial transport. Their work should be consulted for this reason and also because the recent advances in the study of permeation through "inverted soap film" bilayers [85] may eventually help to interpret such measurements in terms of mono-layer properties.

C. Permeability to Ions

The only technique that has been used for this type of measurement is the polarographic method. The problems of applying this technique and of interpreting the results have already been dealt with in a previous section. It should be emphasized that the technique is potentially very versatile, since it can be applied to the study of a variety of charged and uncharged species in both aqueous and nonaqueous media.

The polarographic method has been the subject of several books [46,51, 62], so there is not much point in dealing with the details of experiments, i.e., cleanliness of mercury, the drop time, degassing, etc., in an inevitably superficial manner as part of this chapter. The techniques can be learned from the general references cited, and where special precautions are indicated, the previous discussion on the problems can be consulted for more specific references.

V. CONCLUSION

In this final section the characteristics of monolayer permeation are discussed along with the theories that have been proposed to describe the process. This type of discussion serves to summarize much of what has been learned about monolayer permeation and provides a basis for approaching new problems. In cases where specific new information about monolayer permeation is desired (i.e., a new monolayer and a new permeant), and it is not possible to perform the experiment, the summary in this section allows one to make an educated guess. The summary also points out various experimental and theoretical problems, several of which are discussed in the final paragraphs.

A. Characteristics of Monolayer Permeation

The permeability of monolayers to various substances has been estimated for monolayers composed of various polar and nonpolar groups, at different surface pressures, and at several temperatures. The least permeable monolayers are formed from the saturated, straight chain fatty acids, alcohols, and esters of C_{18} chain length and longer. Monolayers that have been found to be almost freely permeable are those composed of oleic and elaidic acids (which have a double bond in the hydrocarbon chain), cholesterol, and various protein films. In mixed films the resistances appear to add in parallel, that is, the resistance of a binary mixture is lower than that of either component when in a pure film.

The monolayers that are effective diffusion barriers have specific properties which determine the degree of impermeability:

1. The films are of the close-packed condensed type.

2. The surface pressure at which the film becomes relatively impermeable is approximately the surface pressure at which the film becomes incompressible.

3. In a homologous series, the permeability of the compound is related to the length of the hydrocarbon chain.

4. In a series of compounds having the same chain length, the permeability depends on the size of the polar group.

The experimental results also indicate that the permeation process in monolayers differs in many ways from the analogous process in bulk phases [8]:

1. Fick's law is not obeyed in monolayer permeation. Experimentally, the permeation rate is proportional to the concentration difference and area, but not inversely proportional to barrier thickness, as demonstrated by using monolayers of different chain lengths.

2. The magnitudes of the permeability and the activation energy are comparable to analogous values for solids. The monolayer process resembles the permeation of solids, but the variation of the activation energy with the monolayer thickness implies a different kind of process — a one-step, discontinuous process.

3. The differences between the permeabilities of a monolayer to different gases are best explained by the effect of molecular size of the permeating species. The permeation rate appears to vary exponentially with the molecular cross-sectional area of the permeant.

4. The composition and physical state of the monolayer influence the rate of permeation. The monolayer is similar to bulk membranes in this respect, but the permeability appears to be much more dependent on monolayer compressibility.

5. Unlike the case of bulk permeation, there are no interfacial partition effects in monolayer permeation. The permeabilities to the various gases show no correlation with partition coefficients estimated assuming that the monolayer shows the same solubility as hexadecane.

6. In monolayer permeation one observes effects that indicate interference between a permeating gas and another nondiffusing gaseous component of the system.

In more recent work on the permeability of a charged monolayer to ions, it has been shown that the permeability depends on the surface pressure and on the ionic strength [64]. The logarithm of the "permeability" varies linearly with the surface pressure at constant ionic strength, and the slope of the relation shows an order of magnitude change in permeability for a pressure change of about 10 dyn/cm. At constant surface pressure, the permeability increases markedly with ionic strength. These results were obtained for one system only, but they appear reasonable in view of what we know about uncharged permeants and the expected changes in electric repulsion due to variation in ionic atmosphere.

It should be possible to predict the results of new monolayer studies from what we now know about the effects of such factors as permeant size, monolayer polar group size, chain length, and surface phase, etc. The first experiment with a new monolayer substance, the compression isotherm, already gives us a fair amount of information. In general, low compressibility films are less permeable than high compressibility films (e.g., solid films are less permeable than liquid films and, in line with this, permeability decreases as temperature decreases). Other factors, such as the limiting area, can also be determined from the isotherm. These observations are useful guides and also lend support to the theoretical approach which attempts to derive the permeability from the equilibrium compression expansion properties of a monolayer.

B. Theories of Monolayer Permeation

1. The Energy Barrier Theory

Monolayer permeability has been described in terms of an energy barrier at the surface where the monolayer prevents the molecules at the lower end of the kinetic energy distribution from passing through, and therefore acts as an energy sieve. The theory of an activation energy barrier to the evaporation of water through monolayers was first suggested by Langmuir and Langmuir [58] and was developed further by Langmuir and Schaefer[59]. They used a Boltzmann expression to give the fraction f of molecules which have an energy in excess of E, the energy needed to permeate a monolayer.

$$f = \exp(-E/kT) \tag{20}$$

where k is the Boltzmann constant, T is the absolute temperature, and E is the energy needed to compress the monolayer to make a large enough hole. Since the rate of permeation should vary directly as the fraction of molecules having the required energy for penetration, f is proportional to the permeability.

Archer and LaMer [3] considered E in terms of the energy required to form a vacant site in a monolayer lattice, and determined the magnitude of the activation energy for the penetration of the monolayer and of its individual CH_2 groups. The energy barrier for fatty acids in the liquid-condensed (LC) surface phase is about 15 kcal/mole depending upon chain length, and the contribution for each CH_2 group, about 300 cal/mole, generally varies with the lateral compression of a monolayer [16].

The energy barrier theory has proved useful in giving a quantitative description of a permeation process that depends on the surface phase, the length and size of the hydrocarbon chain, the size and nature of the polar group, the surface pressure, and the cross-sectional area of the permeant. However, other properties intrinsic to the monolayer, such as compressibility [74] and free surface area [10], are also needed to describe permeation. In view of the relatively large number of properties that influence E, it has been difficult to visualize a unified physical mechanism for monolayer resistance. Furthermore, the energy barrier formulation has also been unable to account for a number of important observations [10]. The introduction of the "line tension" concept [64] has helped to deal with some of these problems but has also added to the complexity of the model.

2. An Approach via Fluctuations in Monolayer Density

Recently, a different approach to a theory of monolayer permeation has been proposed [10,14]. The new model suggests that free spaces in the monolayer become available for permeation from the natural free area in a lattice and from the equilibrium fluctuations in monolayer density at a (gas molecule-monolayer) collision site. From the entropy change associated with an expansion of a monolayer, it is possible to estimate the probability of a given expansion. The monolayer resistance can then be derived if it is assumed that all local expansions which yield an area equal to or greater than the cross-sectional area of the permeant result in passage through the monolayer. On the basis of these assumptions,

$$r \sim \exp \left[\frac{\gamma (a_0 - A_f)}{kT} \right] \tag{21}$$

where γ is related to the surface tension of a monolayer-covered surface, a_0 is the cross-sectional area of the permeant, and A_f is the free area in a monolayer lattice at a collision site. Although Eq. (21) behaves qualitatively like Eq. (20), the energy barrier is too low. However, if one assumes that the permeant must avoid a collision with the monolayer if it is not to be reflected, the resistance derived on the basis of angular selection predicts an energy barrier that is much closer to observations. The new model also offers a way of considering the extra resistance due to gases that are present at equilibrium (e.g., air when water permeates), since the predicted effect on the measured resistance varies with the concentration of the "inert" gas, as observed [7]. Although this approach requires a much more thorough treatment than has been given to date, it has the advantage of proposing a molecular mechanism that can lead to an understanding of monolayer permeation.

3. The Influence of Subphase Structure

One of the problems that is present in all aspects of monolayer studies is that of the subphase. Since it is impossible to form monolayers without a supporting bulk phase, one can never know to what extent the measured monolayer properties are those of the surface film or of the altered bulk phase surface structure. This problem is particularly perplexing in permeation studies. For fatty acids, about two-thirds of the activation energy barrier has been attributed to the polar group and associated water molecules [3], but oleic acid with the same polar group has no resistance, and molecules with different polar groups (e.g., OH, OC_2H_5) have comparable resistances. While these measurements imply that the subphase structure does not contribute in any large degree to the permeability barrier, there can be no doubt that changes have occurred in the bulk phase surface structure.

C. Current Problems

1. Can Monolayers Increase Mass Transfer Rates?

In 1938, Bull [24] found that the rate of evaporation of water through freshly adsorbed films of ovalbumin, formed on a rotating drum, is greater than from a clean surface. For quiescent surfaces, he found no apparent difference between the two rates. A more recent report [48] claimed enhanced evaporation from quiescent surface films of proteins, but it is not clear from the paper if the conditions were sufficiently controlled to allow the authors to reach this conclusion. In line with these results, it has been suggested [34] that monolayers can increase the rate of evaporation under certain conditions. Most investigators of monolayer permeation do not consider that this is possible, but the problem is worth considering because it relates to very fundamental aspects of monolayer properties.

In a recent paper, Blank and Mussellwhite [18] gave some results that account for Bull's observation by explaining the apparent enhanced evaporation in terms of the greater area of the protein film on the rotating drum. Films which drain slowly cover a wider area for a longer time and evaporate more water even though the per unit area rate is lower. Their experimental evidence obtained for films draining on thin rings shows that a larger ring (where there is a greater possibility of drainage effects) and a slower draining film (such as a protein) both lead to apparent enhanced evaporation rates. In the case of monolayer-coated droplets, where it was previously suggested that there was enhanced transport due to the monolayer, more recent experiments by Snead and Zung [84] show a lower transport rate.

This particular problem will require further work before it is finally resolved; however, at this time the evidence is strongly in favor of the Langmuir view, i.e., that a monolayer resistance can only be positive.

2. Interference Effects

Studies of monolayer permeability to gases [7] have indicated that water vapor is operationally present in the monolayer as an additional resistance. Since the subphase is in equilibrium with the gas phase, water molecules must pass back and forth through the monolayer and they apparently act as a resistance. The movement of water can be reduced by dissolving a salt such as LiCl, and the total vapor pressure can be raised by dissolving a volatile solute such as methanol. These experiments, as well as experiments in which the vapor pressure of water is varied by changing the temperature, all support the view that the vapor molecules from the subphase interfere with the passage of other gas molecules. This result is somewhat unexpected, because two substances can usually interdiffuse independently in an inert bulk phase, and also because foreign substances in monolayers have been shown to increase rather than decrease the permeability [3, 57]. (The latter results do not really conflict, since the foreign substances studied are larger molecules that displace monolayer molecules and thereby create more permeable sites in the film.)

These interference effects should be studied further because they point up an unusual property of monolayer systems that may give important information about physical processes at the molecular level. These effects should also provide some insight into the specific mechanism of monolayer permeation, which still remains to be elucidated.

3. From Monolayers to Bulk Phases

The process of monolayer permeation is essentially a study of the diffusion process at the molecular level. The size of the permeant is comparable to the size of the permeated substance, and the assumption of a multicollision process with consequent averaging, which applies to macroscopic systems, cannot be made in this case. Therefore, monolayer permeation should be of considerable interest to the theoretician, who now has an opportunity to examine some of his assumptions about molecular processes over a wider range. The theoretician also has an additional constraint upon his efforts, since any description of a monolayer process must, upon extrapolation to thicker systems, eventually give rise to the familiar properties of diffusion in bulk systems. It should be added here that natural membrane systems appear to have properties that are related to monolayer systems [13,77] and that monolayers, along with bilayers, offer a convenient model for the study of membrane processes.

Experimentally, a study of the transition from monolayer systems to bulk systems has been initiated by Rose and Quinn [75], who have studied the permeation of multilayer membranes. There are many difficulties in this kind of research (e.g., the inability to maintain a deposition ratio of unity which is so critical to monolayer packing, the problem of cracks in the deposited layers, etc.), but the experimental efforts should eventually give some ideas about the effects of thickness, randomness of orientation, etc. The large volume of experimental work on bilayers [85] also offers the possibility of learning more about the diffusion process in the region of transition between monolayers and bulk phases.

While the results that have been reviewed here indicate that small molecules permeate monolayers with no interaction between permeant and monolayer molecules, there is reason to believe that interactions can occur in other monolayer systems. The well-documented process of "monolayer penetration" has shown that certain surface-active molecules interact with and become incorporated in monolayers [77]. Other kinds of interactions have been found between hydrocarbons and monolayers [26,33]. These studies have not been concerned with transport of the interacting species, so it is not known if the transport rate would depend upon interaction with the monolayer, but there is reason to believe that it would.

These remarks indicate that many of the conclusions about monolayer permeability rest on a limited experimental base, and that further work is needed to investigate the limits and validity of our current knowledge.

Submitted October, 1969

REFERENCES

1.   N. K. Adam and G. Jessop, Proc. Roy. Soc. (London), B98, 206
      (1925).

2.   Per-Ake Albertsson, Partition of Cell Particles and Macromolecules,
      Almqvist and Wiksells, Uppsala, Sweden, 1960.

3.   R. J. Archer and V. K. LaMer, J. Phys. Chem., 59, 200 (1955).

4.   M. K. Baranayev, J. Phys. Chem. USSR, 9, 69 (1937).

5.   G. T. Barnes and V. K. LaMer, in Retardation of Evaporation by Monolayers: Transport Processes (V. K. LaMer, ed.), Academic, New York, 1962, p. 9.

6.   M. Blank, Ph.D. Thesis, Trinity Hall, Cambridge, 1959.

7.   M. Blank, J. Phys. Chem., 65, 1698 (1961).

8.   Ibid., 66, 1911 (1962).

9.   M. Blank, in Retardation of Evaporation by Monolayers: Transport Processes (V. K. LaMer, ed.), Academic, New York, 1962, p. 75.

10. M. Blank, J. Phys. Chem., 68, 2793 (1964).

11. M. Blank, J. Colloid Interface Sci., 22, 51 (1966).

12. M. Blank, in Chemistry, Physics and Application of Surface Active Substances (J. Th. G. Overbeek, ed.), Vol. II, Gordon and Breach, London, 1967, p. 233.

13. M. Blank, J. Gen. Physiol., 52, 191S (1968).

14. M. Blank and J. S. Britten, J. Colloid Sci., 20, 789 (1965).

15. M. Blank and S. Feig, Science, 141, 1173 (1963).

16. M. Blank and V. K. LaMer, in Retardation of Evaporation by Monolayers: Transport Processes (V. K. LaMer, ed.), Academic, New York, 1962, p. 59.

17. M. Blank and I. R. Miller, J. Colloid Interface Sci., 26, 26 (1968).

18. M. Blank and P. R. Mussellwhite, ibid., 27, 188 (1968).

19. M. Blank and F. J. W. Roughton, Trans. Faraday Soc., 56, 1832 (1960).

20. D. P. Boyd and J. M. Marchello, Chem. Eng. Sci., 21, 769 (1966).

21. R. S. Bradley, J. Colloid Sci., 10, 571 (1955).

22. R. Brdicka, Coll. Czech. Chem. Commun., 19, S41 (1954).

23. A. G. Brown, W. C. Thuman, and J. W. McBain, J. Colloid Sci., 8, 508 (1953).

24. H. B. Bull, J. Biol. Chem., 123, 17 (1938).

25. H. S. Carslaw and J. C. Jaeger, Conduction of Heat in Solids, 2nd ed., Oxford University Press, London, 1959.

26. H. D. Cook and H. E. Ries, J. Phys. Chem., 60, 1533 (1956).

27. J. Crank, The Mathematics of Diffusion, Oxford University Press, London, 1956.

28.   P. V. Danckwerts, Appl. Sci. Res., A3, 385 (1952).

29.   J. T. Davies, J. Phys. Chem., 54, 185 (1950).

30.   J. T. Davies, Advan. Chem. Eng., 4, 1 (1963).

31.   J. T. Davies and E. K. Rideal, Interfacial Phenomena, Academic, New York, 1961.

32.   R. B. Dean, Trans. Faraday Soc., 36, 166 (1940).

33.   R. B. Dean, K. E. Hayes, and R. G. Neville, J. Colloid Sci., 8, 377 (1953).

34.   V. Derjaguin, S. P. Bakanov, and Y. S. Kurgin, Colloid J. USSR, 23, 262 (1961).

35.   B. V. Derjaguin, V. A. Fedoseyev, and L. A. Rosenzweig, J. Colloid Interface Sci., 22, 45 (1966).

36.   M. Dixon, Manometric Methods, Cambridge University Press, 1943, p. 20.

37.   I. Fatt, Ann. N. Y. Acad Sci., 148, 81 (1968).

38.   Y. F. Frei and I. R. Miller, J. Phys. Chem., 69, 3018 (1965).

39.   G. L. Gaines, Jr., Insoluble Monolayers at Liquid-Gas Interfaces, Wiley (Interscience), New York, 1966.

40.   H. N. Glazov, J. Phys. Chem. USSR, 11, 484 (1938).

41.   E. A. Harvey and W. Smith, Chem. Eng. Sci., 10, 274 (1959).

42.   J. G. Hawke and A. E. Alexander in Retardation of Evaporation by Monolayers: Transport Processes (V. K. LaMer, ed.), Academic, New York, 1962, p. 67.

43.   J. G. Hawke and A. G. Parts, J. Colloid Sci., 19, 448 (1964).

44.   G. Hedestrand, J. Phys. Chem., 28, 1244 (1924).

45.   S. Heller, Kolloid Z., 136, 120 (1954).

46.   J. Heyrovsky and J. Kuta, Principles of Polarography, Academic, New York, 1966.

47.   J. F. Holliman, J. F. Largier, and F. Sebba, J. Chem. Soc., 1954, 738.

48.   L. K. James and D. J. O. Berry, Science, 140, 312 (1963).

49.   N. L. Jarvis, J. Colloid Sci., 17, 512 (1962).

50.   H. Jeffreys, Proc. Roy. Soc. (London), A118, 195 (1928).

51.   I. M. Kolthoff and J. J. Lingane, Polarography, 2nd ed., Wiley (Interscience), New York, 1952.

52.   J. Koutecky, Coll. Czech. Chem. Commun., 18, 597 (1963).

53.   I. M. Krieger, G. W. Mulholland, and C. S. Dickey, J. Phys. Chem.,
      71, 1123 (1967).

54.   N. Lakshminarayanaiah and A. M. Shanes, Science, 141, 43 (1963).

55.   V. K. LaMer, ed., Retardation of Evaporation by Monolayers:
      Transport Processes, Academic, New York, 1962.

56.   V. K. LaMer, T. W. Healy, and L. A. G. Aylmore, J. Colloid Sci.,
      19, 673 (1964).

57.   V. K. LaMer and M. L. Robbins, J. Phys. Chem., 62, 147 (1958).

58.   I. Langmuir and D. B. Langmuir, ibid., 31, 1719 (1927).

59.   I. Langmuir and V. J. Schaefer, J. Franklin Inst., 235, 119 (1943).

60.   V. Levich, Physicochemical Hydrodynamics, Prentiss Hall, Engle-
      wood Cliffs, N. J., 1962.

61.   M. Linton and K. L. Sutherland, Australian J. Appl. Sci., 9, 18
      (1958).

62.   L. Meites, Polarographic Techniques, 2nd ed., Wiley (Interscience),
      New York, 1965.

63.   I. R. Miller, J. Gen. Physiol., 52, 209S (1968).

64.   I. R. Miller and M. Blank, J. Colloid Interface Sci., 26, 34 (1968).

65.   I. R. Miller, Y. F. Frei, and D. Bach, J. Polymer Sci., 1967, 5A.

66.   I. R. Miller and L. Nanis, J. Colloid Sci., 17, 699 (1962).

67.   V. A. Ogarev and A. A. Trapeznikov, Colloid J. USSR, 28, 544 (1966).

68.   L. Pearson, Ph.D. Thesis, Dept. Colloid Science, Cambridge
      University, 1957.

69.   R. E. Plevan and J. A. Quinn, Am. Inst. Chem. Eng. J., 12, 894
      (1966).

70.   H. M. Princen and S. G. Mason, J. Colloid Sci., 20, 353 (1965).

71.   H. M. Princen, J. T. G. Overbeek, and S. G. Mason, J. Colloid
      Interface Sci., 24, 125 (1967).

72.   Lord Rayleigh, Phil. Mag., 32, 529 (1916).

73.   E. K. Rideal, J. Phys. Chem., 29, 1585 (1925).

74.   H. L. Rosano and V. K. LaMer, ibid., 60, 348 (1956).

75.   G. D. Rose and J. A. Quinn, Science, 159, 636 (1968); J. Colloid
      Interface Sci., 27, 193 (1968).

76.   R. J. Schmidt and O. A. Saunders, Proc. Roy. Soc. (London), A165,
      216 (1938).

77.   J. H. Schulman, B. A. Pethica, A. V. Few, and M. R. J. Salton, Progr. Biophys., 5, 41 (1955).

78.   J. H. Schulman and H. L. Rosano, in Retardation of Evaporation by Monolayers: Transport Processes (V. K. LaMer, ed.), Academic, New York, 1962, p. 97.

79.   F. Sebba and H. V. A. Briscoe, J. Chem. Soc., 1940, 106.

80.   F. Sebba and E. K. Rideal, Trans. Faraday Soc., 37, 273 (1941).

81.   F. Sebba and N. Sutin, J. Chem. Soc., 1952, 2513.

82.   C. I. Sklyarenko and M. K. Baranayev, J. Phys. Chem. USSR, 12, 271 (1938).

83.   C. I. Sklyarenko, M. K. Baranayev, and K. I. Mezyeva, J. Phys. Chem., USSR, 14, 839 (1940).

84.   C. C. Snead and J. T. Zung, J. Colloid Interface Sci., 27, 25 (1968).

85.   H. Ti Tien and A. Louise Diana, Chem. Phys. Lipids, 2, 55 (1968).

Chapter III

ULTRAFILTRATION

Carel J. van Oss

Department of Microbiology
School of Medicine
State University of New York at Buffalo
Buffalo, New York

I. INTRODUCTION

Ultrafiltration is the method by which dissolved molecules are separated from solvent and from other dissolved molecules according to their size by means of a semipermeable membrane, with a pressure differential as the driving force. Other names, such as "hyperfiltration" and "reverse osmosis," have been proposed for this process from time to time. "Hyperfiltration" has little to commend it; it is a Greek-Latin hybrid, it does not indicate anything different from the old and familiar term "ultrafiltration," and thus is best avoided. "Reverse osmosis" is a fairly new description for the same process. It came into fashion in the early 1960's and it more specifically alludes to ultrafiltration when used for the removal of small molecules and ions. It should be clear that no ultrafiltration is possible without applying an upstream pressure that is higher than the osmotic pressure of the solute molecules that are retained by the membrane. Thus, all ultrafiltration is reverse osmosis. However, with even fairly concentrated

89

solutes of a molecular weight of 10^4 daltons or more, the osmotic pressure to be overcome becomes negligible in comparison to the ultrafiltration pressures that are usually applied. For that reason the reverse osmosis aspect of ultrafiltration (which was until 1960 principally applied to macromolecules) has up to recent times rarely (if ever) been stressed; but for the separation of small molecules and ions from solvent by ultrafiltration, the term "reverse osmosis" is, and will doubtless continue to be, widely used. We shall also use it in that sense.

Suspended particles of the order of $0.5\,\mu$ and larger can also be separated from the suspending medium by means of membranes. This process is deemed to be filtration (or "fine filtration" or "microfiltration") and will not be treated here. The molecular weight range of soluble molecules to which ultrafiltration can be applied is as wide as the molecular weight range of soluble molecules: It goes from 10^7 daltons and more down to that of solutes and solvated ions of a molecular weight barely five times higher than that of the solvent molecules. The smallest dissolved nonionized solutes that have been successfully separated from their solvent have a molecular weight of about 10^2. The smallest ion that has been successfully separated from its solvent (water) is lithium [1] (atomic weight 7); its hydrated weight, though not precisely known, is of the order of 10^2 daltons.

Ultrafiltration as a separation method has been practiced for well over a century, but it is only since the early 1960's that, thanks to the new developments in membrane design, it has begun to play the role that has been predicted for it for many years. The older attempts at ultrafiltration suffered from two alternative ills: The flow rates were either terribly slow, or the pore sizes as well as the flow rates were irreproducible in the most frustrating manner.

The first of these ills, the slowness of the attainable fluxes, is inherent in the homogeneous membranes that have been on the market for a number of decades. Cellophane is a classical example of a widely available and reproducible membrane with a reasonably defined pore size (about 2-3 mμ). Its flow rate, with a membrane thickness of 50 μ (when swollen in water), is only about 2 ml/hr/100 cm^2/30 psi. Thus, to obtain any yield at all with such membranes, fairly sophisticated devices that combine high operating pressures with large membrane surfaces have to be used.

The second of these ills, the irreproducibility of membrane pore sizes as well as flow rates, is due to the anisotropy of many membranes. This anisotropy, as will be shown below, can be applied to great advantage to obtain high fluxes and sharp separations. But when this property is unsuspected (as it was until the early 1960's), the most frustrating phenomena will recur. Briefly, an anisotropic membrane has one surface that has a much smaller pore size than the rest of the membrane, due to the process by which it is manufactured. When used with the surface with the smaller pore size ("the skin") toward the liquid that is to be ultrafiltered (or with the skin "up"), relatively small molecules will be retained by the membrane without clogging (or "fouling" or "surface polarization") and at a high flow rate. But when used with the skin "down," the membrane will clog very rapidly by filling up with molecules that can penetrate it more or less deeply, so that the flow rate diminishes very quickly. Nevertheless, this very clogging of the inside of the membrane will eventually allow the passing of some of the molecules that the same membrane, with the skinned side "up," would have stopped completely.

In many cases the side of the membrane that is "skinned" is not visually apparent. Thus, where anisotropy exists but is unsuspected (cryptoanisotropy) [2], membranes will on an average be used with their right side up for only about 50% of the time. Until the fairly recent conscious application of the unique advantages of anisotropy, the properties of such membranes were to all appearances quite irreproducible.

The discovery of anisotropy in membranes [3] and its conscious and systematic application to ultrafiltration have caused an upsurge in publications during the last 10 years [4-6]. Many separation possibilities that hitherto seemed theoretically feasible but for practical reasons remained quite elusive are now being realized.

II. MEMBRANES

A. Homogeneous Membranes

There is no longer any particular advantage in using homogeneous membranes in ultrafilters. The most universally available membrane, cellophane (made of regenerated cellulose), with a flow rate of 2 ml/hr/100 cm^2/30 psi [7], is about 100 times slower than an anisotropic membrane with the same pore size (2-3 mμ), when actually retaining proteins from a solution [8]. To obtain any yield at all with cellophane, quite large membrane surface areas as well as high operating pressures of the order of at least 300 psi are indispensable. The pore size of cellophane can be modified, within rather narrow limits, by chemical [7,9] as well as mechanical means. Among the latter methods it is interesting to note that unidimensional stretching decreases the effective pore size, while bidimensional stretching increases it [10].

The most complete review on other homogeneous ultrafiltration membranes, from the earliest beginnings until 1936, was compiled by Ferry [11]. More recent developments have been discussed by Friedlander and Rickles [12] in 1965 and by Rickles [13] in 1967, and the most recent review on ultrafiltration membranes appeared in 1971 [2].

Homogeneous membranes can now also be prepared in the shape of extremely fine capillaries. DuPont manufactures them from nylon [2], mainly for reverse osmosis, and Dow makes capillaries from cellulose triacetate that seem promising for use in artificial kidneys [14]. For ultrafiltration, capillary membranes are best used by applying the pressure from the outside, the ultrafiltrate flowing off inside the tiny tubes. They need no support and, although the flow rates attainable per unit membrane area are slow, enormous areas of capillary tubing can be stored in small volumes. Recently, the use of an ice membrane has been proposed, which seems to have some promise as a salt-retaining membrane, although no precise data on its effectiveness were given [15].

B. Anisotropic Membranes

Anisotropic or "skinned" membranes are formed under one or both of the following conditions:

1. Upon its immersion into the coagulating liquid, the top layer of the cast solution coagulates first. The coagulating liquid then has to pass the newly created top skin, which slows down its further penetration into the cast solution and gives rise to a more porous structure farther down.

2. If, prior to gellification (or coagulation), superficial evaporation of solvent from the cast solution is allowed to take place, this will also result in a top layer that is more concentrated than the rest of the membrane's bulk.

As a general rule, an anisotropic membrane must always be used with that side "up" that was the first to touch the coagulating liquid upon its formation, or with that side "up" from which the strongest initial evaporation took place prior to gellification or coagulation, or both. (In both cases that is, of course, the same side.)

Anisotropic membranes were first made by Loeb and Sourirajan [3]. These membranes were quite dense and were intended for the retention of salt ions from seawater (reverse osmosis) and they were indeed the first membranes that made reverse osmosis practicable. More recent reverse osmosis membranes have been described by Loeb [16], and a general review of reverse osmosis has been given by Lonsdale [17].

The first anisotropic membranes deliberately made for stopping proteins and other water-soluble polymers in order to concentrate them were made by Michaels and his collaborators from polyelectrolyte complexes [18]. These membranes are now marketed by the Amicon Corporation in Lexington, Massachusetts, under the name Diaflo membranes. There exists a much older anisotropic protein-stopping membrane in the shape of a collodion bag (made of cellulose nitrate with a skin on the high pressure side, manufactured by Sartorius and marketed by Brinkman Instruments in Westbury, New York). These membranes are still quite popular for the concentration of small volumes of protein solution. The particular method of their manufacture in the shape of collodion bags [18a] guarantees that the skin is always on the inside, so that although these membranes had already been manufactured for generations before the discovery of anisotropy, their geometry precludes accidental improper use. Another useful anisotropic protein-concentrating membrane (made of cellulose acetate) is marketed by the Gelman Instruments Company at Ann Arbor, Michigan and called a protein exclusion membrane ("P. E. M. "). Two other commercially manufactured membranes that are marketed for the purpose of concentrating protein solutions are less satisfactory because they have been found to leak protein. These are the "Pellicon" membranes (manufactured by Millipore in Bedford, Massachusetts) and the Schleicher & Schuell (Keene, New Hampshire) "LSG" protein-concentrating membranes. The "LSG" membranes only leak when they are used to concentrate protein at concentrations of 1% and over [4, 8].

Anisotropic protein (and other water-soluble polymer) -concentrating membranes are relatively easy to manufacture in the laboratory. The easiest material to use for reproducible results is cellulose diacetate. The method of preparation of these membranes has been described in various places [4, 8, 19], but for the sake of completeness it will be given again below.

Twenty-five g of cellulose acetate (39.8% acetate), ASTM viscosity 3 (Eastman #4644), are slowly added under constant stirring with a pestle to 75 ml of acetone and 50 ml of formamide in a mortar of about 130-mm opening. Stirring is continued until all lumps are dissolved. Alternatively, the mixing process utilizes a Servall Omnimixer (Ivan Sorvall, Inc., Norwalk, Connecticut). Since in the latter case the use of a closed chamber cuts down the solvent loss (primarily of acetone), the initial acetone level is reduced from 75 to 65 ml, so that the proportions in the final mixture are the same as in the conventionally prepared mixture.

The mixture is then poured into a 500-ml filtering flask which is placed in a 55°C water bath. About 40 cm Hg vacuum is applied to the flask until all bubbles disappear. Approximately 15 ml of the mixture are then poured in a ribbon across the short side of a 17 X 25-cm glass plate (having metal runners 0.15 mm thick along its two long edges). The mixture is then quickly and evenly spread over the glass plate with the help of a 15-cm-long glass tube by drawing the tube in one long smooth horizontal motion along the runners. An excess of about two-thirds of the applied mixture is swept off the plate and discarded. The glass plate with the spread-out mixture is immersed as quickly as possible in an ice water bath and kept for at least 1 hr before the membrane is lifted from the plate.

The almost instantaneous coagulation of the upper surface of the mixture upon its immersion in cold water causes the formation of a very thin skin which is denser than the rest of the membrane structure. This is the actual protein-stopping skin, and care must always be taken to use the membrane with the side up that was away from the glass plate when it was formed. As it is well nigh impossible to distinguish the top of a wet membrane from its bottom, it is advisable to mark one edge of every newly made membrane with an asymmetric perforation pattern, so as to record the direction of its anisotropy [19]. Figure 1 clearly shows the enormous difference in pore size between the skin and the rest of the mass of this membrane.

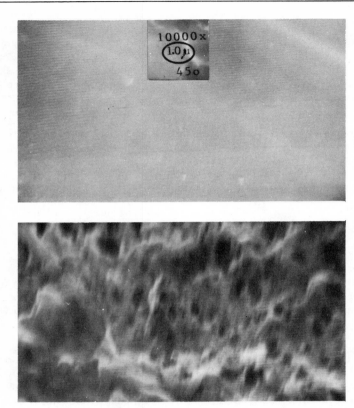

Fig. 1. Scanning electron micrograph of the protein-retaining membrane. Top: "skinned" top of the membrane (approximately 1 μ thick).
Bottom: porous bottom of the membrane. (X10,000. From Ref. [8].)

The cutoff of this membrane (that is, the smallest molecular weight of
globular polymer that it retains completely) is about 20,000 daltons. To obtain a membrane that will concentrate polymers of a molecular weight of
about 6000 daltons, the same recipe can be applied, using 35 ml of formamide
instead of 50 ml. The influence of the thickness of the membrane, the pressure applied on the liquid to be ultrafiltered, and the concentration of protein
in the solution to be ultrafiltered have been described in detail [8]. Although
the best flow rates can be obtained with these membranes when they have
been continuously kept in water from the moment of their manufacture, these
membranes can be dried, provided that this is done after either soaking them
for about 1 h in a 50% glycerol-water solution [8] or soaking them in a 0.1%
sodium dodecyl sulfate solution [8,20]. Prior to their use such dried membranes should of course be washed very thoroughly.

The flow rate of the above-described membrane, when stopping about
1% protein in solution, is 350 ml/h/100 cm^2 of membrane/30 psi pressure.

With more dilute protein solutions these flow rates can be as much as doubled. Generally speaking, drying the membrane by either of the two methods described above will result in a reduction in flow rate of 10-20%. The considerably tighter membrane mentioned above has a flow rate of about 50% of that of the wider-pored membrane.

A new membrane for the fractionation of proteins which has a cut-off at a molecular weight of approximately 300,000 daltons has been described recently [4]. One of its drawbacks is that, although molecules of a molecular weight of over 300,000 daltons are integrally removed from the ultrafiltrate and the other proteins of smaller molecular weight are present in the same proportion in which they were present in the original solution, the total portion of the protein that passes the membrane at all is very low, of the order of 10-20% at the most. It is becoming apparent that the presence of fairly large quantities of other proteins of varying molecular weight can play an important role in the retention or nonretention of other proteins of a given molecular weight and also tends to influence the final protein concentration in the ultrafiltrate.

The most important use of anisotropic polymer-retaining membranes within the whole range of sizes of biopolymers of varying molecular weight (anywhere from 10,000 to several million daltons) is for the rapid concentration of solutions of these biopolymers with the least possible denaturation. With the help of anisotropic membranes, it is now quite feasible to concentrate very dilute protein solutions (when available in quantities of about 1 liter) as much as 500 times in only a few hours [8,19].

It was recently noted by Kopeček and Sourirajan [21] that anisotropic cellulose acetate reverse osmosis membranes acquire a significantly higher flow rate (more than 20% higher), without increase in effective pore size, when they are first used upside down to ultrafilter distilled water, under normal operating pressures; they call this "back-pressure" treatment. The present author [22] found that this effect can also be produced when protein-stopping or protein-fractionating membranes undergo such treatment. The exact reason for this improvement in flow rate is not quite clear; in our opinion it is likely to be related to an increase in pore size of the porous bulk of the membrane, because the pore size of the "skin" of the membrane is apparently not affected. Kopeček and Sourirajan, on the other hand, believe that the back-pressure treatment mainly acts on the "skin" of the membrane and that it has a more pronounced widening effect on the smaller pores than on the wider pores [21].

Another sort of anisotropic membrane which, however, cannot be properly called a skinned membrane, has recently been developed by Marcynkowsky, Kraus, Johnson, Shor, and their collaborators [23-25]. This membrane is made by depositing minute colloidal particles on a very porous support whose pores are too small for the particles to filter through; the membrane is dynamically formed when the particles, instead of filtering through the support, are deposited on top of it. Colloidal particles used are, for instance, zirconium hydroxide, organic polyelectrolytes, and even humic acid. These membranes have been used mainly for reverse osmosis. The colloidal particles have to be added continuously on the high pressure side, and with this type of membrane, even more than with the ones described above, continuous turbulence is of great importance in preventing surface polarization. The effective pore size attainable with these membranes seems to be somewhat larger than that of the densest anisotropic cellulose acetate membranes, as their salt rejection is much less complete.

Finally, the present author has described [7] the use of animal membranes, which either have some analogy with the dynamically formed membranes that are described above, or are of a new and different kind altogether. As they are definitely not homogeneous, they will be described here. They are tubular membranes, consisting of the cleaned small intestines of Chinese pigs, which have been preserved in salt. Once the salt has been leached out with water and the tubular membranes have been opened up, the blown-up wet membranes are mildly air-dried and are ready for use. These membranes are exceedingly pressure sensitive, and one can make use of this property to obtain a range of different pore sizes by simply varying the ultrafiltration pressure, in a perfectly reversible and reproducible manner (see Table I). Remarkably, the flow rate of these membranes remained the same at all applied pressures (50 ml/100 cm^2/h), so that it must be supposed that the pore diameters decrease as the membranes get more compressed and that what would have been gained in flow rate, due to the shortening of the pores, is counterbalanced by the simultaneous shrinkage of their diameters [7].

For a variety of practical reasons these animal membranes are not very likely to be widely used for the separation of viruses and other very large molecules. But the very fact of the existence of membranes whose pore size can be controlled by the ultrafiltration pressure should provide encouragement for attempts to synthesize similar membranes from more homogeneous materials.

These animal membranes have never been used with liquids that were completely devoid of suspended material, so it is possible that their actual ultrafiltering component could have been dynamically formed by the suspended material [23-25].

C. Hydrophilic and Hydrophobic Membranes

Without a fairly elaborate pretreatment, hydrophilic membranes cannot be used for the ultrafiltration of hydrophobic liquids that do not wet them. If the initial pore size as well as the initial flux is to be retained, a hydrophilic membrane has to be wetted with water and then the water must be gradually replaced in small steps by more and more hydrophobic solvents before a totally hydrophobic solvent can be successfully ultrafiltered through the membrane. For instance, a water-swollen hydrophilic membrane can be transformed into a hydrophobic solvent ultrafiltering membrane by first replacing

TABLE I

Variability of Pore Size of Intestinal Membrane with Pressure

Pressure, psi	Approximate pore size, mμ	Impermeable to	Permeable to
5	100	Vaccinia virus (~250 mμ diameter) [26]	Foot and mouth disease virus (~30 mμ diameter) [27]
30	25	Foot and mouth disease virus (~30 mμ diameter)	Serum proteins (~10-20 mμ)

the water with a 50% water-ethanol mixture, then slowly increasing the ethanol content of successive baths up to 100%, then gradually replacing the alcohol that now permeates the membrane with, for instance, benzene, and finally replacing the benzene with the desired hydrophobic liquid [7]. The smaller the pore size of the membrane, the more the precaution of gradual replacement of hydrophilic by hydrophobic solvents is necessary, due to the increased pressures in pores of small radii caused by the surface tension, according to Laplace's law [2].

For the same reason, one has to be very careful when drying membranes, particularly membranes of the smaller pore sizes. Generally speaking, there are two methods by which membranes with even small pore sizes can be dried. The first method consists of imbibing the membrane with a plasticizing liquid that does not evaporate spontaneously, such as glycerol or glycol. Although the water can be removed by evaporation, another liquid will remain to prevent the membrane from drying out and shrinking. The second method is to treat the membrane with an aqueous solution of a surface-active agent prior to drying [20]. The surface tension will then generally be sufficiently decreased for the membrane to be dried without undergoing irreversible shrinkage of its pores. (It goes without saying, of course, that membranes can only be rendered useful for ultra-filtering those organic solvents in which they are not themselves soluble.)

Not much work has been done up to now with hydrophobic ultrafiltration membranes. Most of the published work centers around Sourirajan's publication [28]; see also [2].

III. ULTRAFILTRATION

A. Analytical Uses of Ultrafiltration

1. Determination of Molecular Weights

The attempt to determine molecular weights of macromolecules and viruses was one of the earliest uses of ultrafiltration. Much work on the preparation of membranes of different pore sizes was done by Elford [29] and Grabar [30] in particular (see Ferry [11] for a complete review of this work). Nevertheless, ultrafiltration has never become an important method for the determination of molecular weights. This is doubtless partly owing to the above-mentioned cryptoanisotropy [2], but it is also certainly due to the emergence at about the same time, that is in the 1920's and 1930's, of much more precise methods for determining molecular weights by analytical ultracentrifugation [31] and more recently by electron microscopy. In any case, ultrafiltration is and will remain fundamentally a preparative method. As such, it can be most useful as a preliminary step in analysis, but it will never replace purely analytical methods of great precision. The same principles apply to other preparative methods, such as gel filtration [32] or certain varieties of gel electrophoresis [33]. These methods, though useful for determining rough orders of magnitude of molecular weights, are not ultimately analytical methods.

2. Measurement of Binding Forces

Ultrafiltration can serve in a quite unique fashion for the measurement of binding forces and for the measurement of binding in general between small and much larger molecules or ions. The binding between small and larger molecules can always be measured by ultrafiltration as long as a membrane can be found that is impermeable to the larger molecule but totally or at least largely permeable to the smaller molecule. This method is extremely simple and yields many data in a single experiment. Briefly, when a known amount of small molecules or ions is added to a solution of large molecules and when this solution is then ultrafiltered through a membrane that passes only the small molecules or ions, the simple measurement of the concentration of small molecules or ions found in the ultrafiltrate, when compared with the total amount of small molecules or ions added to the original solution, will, by difference, yield the amount of small molecules or ions that has remained bound to the macromolecules that are retained by the membrane. To account for any small molecules that may have been retained by the membrane, a parallel experiment can be run using solutions that are identical in their composition of microsolutes but which lack the macromolecules. A method has been worked out by which the tendency of certain membranes to retain a small proportion of the small molecules or ions can be taken into account [34]. The present author has done a number of determinations of the sodium and potassium binding of a variety of proteins and also of proteins associated with a number of different surface-active agents. In particular, the combination of bovine serum albumin (BSA) with quaternary ammonium bases showed

strong sodium binding and hardly any potassium binding, while in combination with soap (mainly Na-stearate and palmitate), on the other hand, the
BSA preferentially bound potassium [35]. For such experiments it is essential that the inside of the ultrafilter be well stirred, as clogging of the membrane and other undesirable phenomena may otherwise occur. It has been
shown that in equilibrium situations, e.g., in the simple dialysis of the same
substances, this preferential binding does not occur [35]; for this reason,
the multiple ultrafiltration results published by Salminen [38] cannot be regarded as very conclusive, since no stirring or other agitatory precautions
were taken in any of the ultrafiltration experiments described by him. More
recent work on ultrafiltration as a method for measuring binding has been
published by Blatt and co-workers [39] and also by Paulus [40]. Paulus has
devised an ultrafilter in which multiple ultrafiltrations of this type can be
done at the same time. This method suffers, however, from the same flaw
as the one mentioned in Salminen's work [38]: **there is no stirring of the**
liquid to be ultrafiltered, so that all sorts of side effects can and probably do
obtain.

B. Preparative Uses of Ultrafiltration

1. Concentration of Solutes

Concentration of solutes of large and, nowadays, even small molecules
has become one of the most important functions of ultrafiltration as a preparative method. And ultrafiltration has become one of the most important
preparative methods because, auxiliary to numerour other methods of separation and purification, it opens up the possibility of reconcentrating all
fractions that have become too dilute to allow further purification. Thus,
ultrafiltration as a preparative method has its most important function as an
intermediate but increasingly indispensable step in a variety of separation
procedures. The use of anisotropic ultrafiltration membranes for the concentration of proteins has been discussed in detail above.

Reverse osmosis also falls under this heading and has been mentioned
for the concentration of dilute sugar juices [41].

2. Removal of Solutes

Ultrafiltration is important in a variety of fields for removing macromolecules or even smaller molecules from solutions, for a great number of
different purposes.

a. In clinical chemistry, it is frequently necessary to do various determinations on deproteinized sera or other body fluids. Ultrafiltration is play-
ing an increasingly important role here [42].

b. For the removal of viruses, ultrafiltration can have a very useful
role. Membranes can now be made that have a small enough pore size to be
impermeable to all the known viruses while remaining permeable to practically all, if not all, of the plasma proteins [4]. Thus, at least theoretically,

plasma can be ultrafiltered and obtained free of all possible viruses, including hepatitis virus. (One of the main drawbacks to the practicable use of this kind of method is that concentrated protein solutions that have to be ultrafiltered through membranes with pore sizes not far removed from the size of the proteins themselves, ultrafilter with extremely low efficiency. That is, although all proteins of a given size will be able to pass the membrane, the concentration of these proteins in the ultrafiltrate will remain very low.)

c. The removal of pyrogens and antigens from injectable pharmaceuticals is a fairly important application of ultrafiltration [43]. It is possible to make injectable hydrolysates of a variety of protein solutions and simply to ultrafilter them through a membrane that is impermeable to all proteins, thus obtaining an injectable hydrolysate which will not contain any antigens large enough to elicit antibody formation nor any pyrogens that will give rise to undesirable febrile side effects.

d. Ultrafiltration is being studied in a number of laboratories for possible use in artificial kidneys. In such apparatus part of the blood stream of a patient is ultrafiltered and the protein-free ultrafiltrate is either purified of all undesirable constituents such as urea, creatinine, etc., and then reintegrated into the circulation, or the ultrafiltrate is simply discarded and replaced by nonnoxious solutions.

e. Finally, one of the large-scale applications of ultrafiltration that has to be mentioned is, of course, reverse osmosis. Desalination of brackish water or seawater is nothing but the removal of undesired solutes from solvents [16,17,44]. Apart from desalination, the removal of other solutes is frequently desirable; and at the moment, reverse osmosis is much studied for that purpose in the field of waste water purification [45,46]. Ultrafiltration through somewhat more porous membranes is also being considered[47].

3. Separation of Molecules According to Size

a. Influence of the Degree of Asymmetry of Molecules on Their Ultrafiltrability. Relatively little work has been done on this aspect of ultrafiltration of macromolecules, but it is obvious that very long and thin molecules of a given molecular weight will be able to pass pores of a given size much more easily than more spherical molecules of the same molecular weight. This was first noticed by the present author when mixtures of dissolved starch molecules of two different shapes were ultrafiltered [7]. The long, unbranched amylose molecules of fairly high molecular weight ultrafiltered more easily than the much more compact and branched amylopectine molecules of a comparable molecular weight. A table published by the Amicon Corporation [6, 48], showing the ultrafiltrability of a number of biopolymers of increasing molecular weight through a variety of different membranes, also shows a remarkable gap for dextran of molecular weight 110,000. This linear polymer, which figures in the middle of a series of much more symmetrical molecules, ultrafilters in significant quantities through membranes that are totally impermeable to the much smaller symmetrical molecules. These facts form another argument for not characterizing the porosity of a

membrane by the molecular weight of the substance that it is just capable of
stopping, but of indicating the porosity by the molecular weight and shape of
the molecules it can retain. To that purpose it is always very useful when
doing ultracentrifugal studies on macromolecules to determine sedimentation
as well as diffusion coefficients, in order to learn the molecular weight as
well as the asymmetry factor of the macromolecule in question [31].

   b. Separation of Viruses from One Another. There has not been a tre-
mendous amount of published work on this subject in recent times. One
observation was that of the present author, who separated vaccinia virus from
foot and mouth disease virus using an animal membrane [7,49]. Homogeneous cellulose ester membranes have been used by Cliver for the separation
of a number of viruses [50]. Generally speaking, for the separation of substances of high molecular weight (or of a fairly large size), such as viruses,
it would not seem to be of great importance to use anisotropic membranes.

   c. Separation of Macroglobulins from Globulins and Smaller Proteins.
Because of the different functions of the blood macroglobulins (which have a
molecular weight of about 860,000 daltons), the other blood serum globulins
(molecular weight, 160,000), and albumin (molecular weight, 69,000), there
exists a real need for an easy separation of those classes of proteins. The
first attempt to separate the macroglobulins from the other proteins was
made by the present author with an anisotropic membrane made of agarose
[51]. This membrane successfully separates the macroglobulins from the
other serum proteins, but its thickness (several millimeters) allows only very
slow flow rates, which makes it rather impractical. More recently, the present author and his collaborators have succeeded in making a thin anisotropic
membrane of a mixture of cellulose nitrate and cellulose acetate which stops
all 19 S macroglobulins from serum and passes 7 S globulins and albumin.
The preparation of the membrane has already been published elsewhere [4,
52] but is given here for the sake of completeness:

   Fifty ml of glacial acetic acid and 40 ml of acetone are mixed.
To this mixture are added 15 g of cellulose nitrate (30% ethyl alcohol, viscosity 16.2, A.B., Grade DHB 14 E, DuPont). (The principal role of the cellulose nitrate is to lend more mechanical
strength to this fairly open-pored membrane.) The whole is then
mixed further until uniform; then 15 g of cellulose diacetate (39.8%,
ASTM 3, Eastman No. 4644) and 40 ml of formamide are added.
Mixing is resumed until the mixture is uniform. From this point,
the process continues as described in Sec. II.B for the preparation
of cellulose acetate protein-stopping membranes.

   The mixture is cast in the same way as described in Sec. II.B.,
except that 0.25-mm-thick runners are used and the casting is performed on stainless steel instead of glass plates to prevent the
membranes from sticking to the plates. Once drawn, the membrane
is "superskinned" with the aid of an air blast at 20°C, provided by a
hair dryer held 15 cm above the membrane for 60 sec. Then the

membrane is immersed in an ice water bath for at least 1 h prior to use. This membrane may be dried after immersion in 50% glycerol; however, due to its strong anisotropy and its brittleness in the total absence of plasticizer, treatment with 0.1% sodium dodecyl sulfate cannot be used as a pretreatment for drying.

With this membrane the optimal pressure is 10 psi, but pressures up to 20 psi can be used. The actual "skin" is about $0.3\,\mu$ thick, as judged by scanning electron microscopy.

The flow rates with this membrane are about 150 ml/h/100 cm^2/10 psi for 1% protein and 60 ml/h when ultrafiltering whole serum.

A drawback of this type of membrane is that, although it stops all macroglobulins and although the smaller proteins that pass the membrane are passed in the proportion in which they are normally present in normal serum, the total yield is fairly low: Normally only 10-20% of the total proteins pass the membrane at all. Nevertheless, for a number of analytical purposes this membrane serves a useful purpose; the important aspect is that the ultrafiltrate will be devoid of macroglobulins, while albumin and the 7 S globulins are present in close to normal proportions.

d. Separation of Globulins from Albumin. The present author and his collaborators have succeeded in making an anisotropic membrane from cellulose acetate according to the recipe for a protein-stopping cellulose acetate membrane, but using 100 ml of formamide per batch instead of 50 ml [8]. Such a membrane will indeed retain 7 S human gamma globulin of molecular weight 160,000 while passing human serum albumin of molecular weight 69,000, but it will only do so if these proteins are present in extremely low concentrations, that is, at a total protein concentration not exceeding 0.1%. When whole serum, diluted to that low a concentration of protein, was ultrafiltered, no separation could be achieved, for unexplained reasons. The Amicon Corporation markets a membrane which allegedly can separate albumin from globulin, again at low concentrations; however, we found that even with the artificial mixture from which our membranes can separate these two, the Amicon membrane will not do so, but passes an unseparated mixture of diluted protein. Only when pure albumin in very dilute form is present alone, will the Amicon membrane pass it; and it will stop pure human gamma globulin only if that is present alone in dilute form. These results show that to stop one protein by itself and to pass another, smaller protein by itself is not at all the same thing as to separate the two proteins when they are present together in a mixture. These results also show that it is essential to monitor such experiments with the only analytical method that is capable of giving molecular weight distributions of the mixtures of the biopolymers in solution that have been put to ultrafilter, as well as of their ultrafiltrates, i.e., analytical ultracentrifugation.

e. Separation of Proteins of a Molecular Weight Smaller than 70,000 from one Another. Not much work has been done in this particular field, but membranes that separate smaller proteins from one another are relatively

easy to make, in the same manner as our protein-stopping anisotropic cellulose acetate membrane [8] but with 35, 30, 25, or fewer milliliters of formamide per batch instead of 50 ml. These amounts form a series of membranes
of decreasing pore sizes which are capable of fractionating molecules of
10,000 to only a few thousand daltons molecular weight.

IV. ULTRAFILTERS

A. Laboratory Ultrafilters

It has been noted by McBain [53], Erschler [54], and Trautman and
Ambard [55,56] that fouling (or clogging) of ultrafiltration membranes by
retained macromolecules as well as other types of surface polarization caused
by salt accumulation can be largely prevented by constantly agitating the liquid
above the membrane. Initially, this was always done by shaking or oscillating
the entire ultrafilter. In the 1950's magnetic stirring devices became generally available, which permitted the development of the modern simple and
efficient laboratory ultrafilter (see Fig. 2). (However, when an ultrafilter is

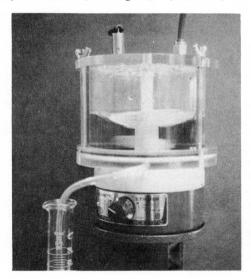

Fig. 2. Photograph of a laboratory ultrafilter cell. Compressed nitrogen is introduced into the cell through the tube at the upper right. At the
upper left is a safety valve, shown in the closed position. The entire cell is
placed on a magnetic stirrer which activates the magnetic stirring rod inside
the cell. It is clearly visible how this causes the liquid to rotate above the
membrane. The ultrafiltrate emerges through the tube in the bottom of the
cell (in the lower middle of the picture) and is collected in a graduated cylinder (lower left). The simplicity of construction of the cell and the advantage
of the transparent cell housing are obvious from this picture.

used in an electrochemically sensitive system, one must not lose sight of the fact that a magnetic stirring device creates a rotating magnetic field close to the plane of the membrane, which gives rise to an electric field across the membrane that can have considerable effects [57].)

Figure 2 shows an example of the most practical type of laboratory ultrafilter. The pressure chamber of the ultrafilter is best made of transparent plastic because it is of inestimable value to be able to see how much liquid still remains in the ultrafilter. To date, the only commercially available ultrafilters of this type are made and marketed by the Amicon Corporation, Lexington, Massachusetts and by the Gelman Instrument Company, Ann Arbor, Michigan. The pressure source for this type of ultrafilter is compressed gas, usually nitrogen.

When no source of compressed nitrogen is available, and if no great importance is attached to the possibility of stirring the solution that is to be ultrafiltered, a much simpler device may be used, in the form of a simple centrifuge filter holder. There is one type, marketed by Amicon, which is made to be used with conical membranes (also made by Amicon). For flat membranes there is an excellent filter holder that fits most tabletop centrifuges, made by the Millipore Corporation, Bedford, Massachusetts [52], and another one, which houses a membrane of rather smaller diameter, made by the Gelman Instrument Company.

B. Larger Ultrafilters

In large-scale ultrafiltration, the prevention of surface polarization is even more crucial than it is in the laboratory. It was first noticed [58] by the present author that with a tubular membrane system, the maximum reduction of surface polarization was obtained when a turbulent regime (at Re = 2300) was reached (see also Ref. [7]). Higher Reynolds numbers did not cause a further decrease in surface polarization when ultrafiltering protein solution, as judged by flow rates and protein retentions [7]. With somewhat lower Reynolds numbers, turbulence in the vicinity of the membrane, resulting in a strong decrease in surface polarization, could still be obtained with the help of turbulence-promoting ridges, pressed into the membrane by the special pattern of the porous support woven of monofilament saran thread [7,58].

In the closest vicinity of the membrane, even in turbulent regimes, there is always a nonturbulent and virtually stationary boundary layer. But that boundary layer will always be of lesser thickness when the bulk of the flow is turbulent than with laminar flow. This is an important consideration in the design of large-scale ultrafilters or reverse osmosis plants. (The higher the product flow rate through the membrane, the more important it is to have a minimally thin boundary layer [59,61]).

The following types of large-scale ultrafilters have been described:

1. Tubular Systems

These are very similar to [62], or based upon, the system devised by
the present author [7,58]. The advantages are that turbulence over the whole
membrane area is easily attained, high pressures can be supported by relatively thin tubes, and leakage does not readily occur and when it occurs it can
be easily corrected. The drawbacks are a relatively high volume of the in-
stallation for a given membrane area and a relatively high volume of circulating liquid. The optimal tube diameter is likely to be between $\frac{1}{4}$ and 2 in. At
smaller diameters the head losses due to friction will become too high with
turbulence, and at larger diameters the volume of circulating liquid becomes
too enormous. In order to increase their efficiency, turbulence-promoting
devices have been proposed for tubular systems [7, 58, 63].

2. Flat Plate Systems

These are more or less similar to filter presses. These systems can
house fairly vast membrane surfaces in a compact volume. The drawbacks
are that turbulence is not easily obtained over the entire membrane area, and
relatively enormous lengths of pressure-tight gaskets per unit membrane
area have to be used, unless membranes can be applied seamlessly by integrally coating a porous plate manifold with them, as seems to have been the
technique used for the apparatus described by Weissman et al. [47].

3. Swiss Roll Systems [64,65]

These systems can also house large membrane surface areas in relatively small volumes, but turbulence is very difficult to attain.

4. Capillary Systems

These systems can house the maximum membrane surface area in a
minimum volume [2], but it is not possible to work with turbulence. On
the other hand, such enormous membrane areas can be accommodated that one
can probably afford low product flow rates per unit membrane area, which
makes turbulence less crucial.

The systems most likely to be the most efficient and to emerge ultimately
in the greatest variety of applications are the tubular and capillary systems.
In the first case, this is due to the ease with which turbulence can be obtained,
and in the capillary system, because these systems are the least dependent
on turbulence. Nevertheless, in at least one kind of flat plate system turbulence seems to have been provoked with apparent ease [47], which may also
augur well for the success of this type of arrangement.

ACKNOWLEDGMENT

This work was supported in part by Public Health Service Research
Grant GM 16256 from the National Institutes of Health.

REFERENCES

1.   A. S. Michaels, H. J. Bixler, and R. M. Hodges, J. Colloid Sci.,
     20, 1034 (1965).

2.   C. J. van Oss, in Progress in Separation and Purification (E. S.
     Perry and C. J. van Oss, eds.), Vol. III, Wiley, New York, 1971,
     p. 97.

3.   S. Loeb and S. Sourirajan, Advan. Chem. Ser., 38, 117 (1962).

4.   C. J. van Oss and P. M. Bronson, in Membrane Science and Tech-
     nology (J. E. Flinn, ed.), Plenum Press, New York, 1970, p. 139.

5.   W. F. Blatt, Am. Lab., March 1969, p. 131.

6.   G. J. Fallick, Process Biochem., 4, 9 (1969).

7.   C. J. van Oss, Dissertation, Paris, 1955; L'Ultrafiltration, de Bussy,
     Amsterdam, 1955.

8.   C. J. van Oss and P. M. Bronson, Separation Sci., 5, 63 (1970).

9.   L. C. Craig, in Analytical Methods of Protein Chemistry (P. Alex-
     ander and R. J. Block, eds.), Vol. I, Pergamon, New York, 1960,
     p. 104.

10. L. C. Craig and W. Konigsberg, J. Phys. Chem., 65, 166 (1961).

11. J. D. Ferry, Chem. Rev., 18, 373 (1936).

12. H. Z. Friedlander and R. N. Rickles, Anal. Chem., 37 (8), 27A
     (1965).

13. R. N. Rickles, Membranes, Technology and Economics, Noyes
     Development Co., Park Ridge, N. J., 1967.

14. R. D. Stewart, E. D. Baretta, J. C. Cerny, and H. I. Mahon,
     Invest. Urol., 3, 614 (1966).

15. R. D. Miller, Science, 169, 584 (1970).

16. S. Loeb, in Desalination by Reverse Osmosis (U. Merten, ed.), M.I.T.
     Press, Cambridge, Mass., 1966, p. 55; U. S. Pat. 3,133,132 (1964).

17. H. K. Lonsdale, in Progress in Separation and Purification (E. S.
     Perry and C. J. van Oss, eds.), Vol. III, Wiley, New York, 1971,
     p.

18. A. S. Michaels, Ind. Eng. Chem., 57, 32 (1965).

18a. J. Duclaux, Ultrafiltration, Vol. I, Hermann, Paris, 1945, pp. 19-26.

19. C. J. van Oss, C. R. McConnell, R. K. Tompkins, and P. M. Bron-
     son, Clin. Chem., 15, 699 (1969).

20. K. D. Vos and F. O. Burris, Ind. Eng. Chem. Prod. Res. Develop.,
     8, 84 (1969).

21. J. Kopeček and S. Sourirajan, J. Appl. Polymer Sci., 13, 637 (1969).

22. C. J. van Oss and P. M. Bronson, unpublished results.

23. A. E. Marcynkowsky, K. A. Kraus, H. O. Phillips, J. S. Johnson, and A. J. Shor, J. Am. Chem. Soc., 88, 5744 (1966).

24. K. A. Kraus, A. J. Shor, and J. S. Johnson, Desalination, 2, 243 (1967).

25. A. J. Shor, K. A. Kraus, J. S. Johnson, and W. T. Smith, Ind. Eng. Chem. Fundamentals, 7, 44 (1968).

26. S. Dales, J. Cell Biol., 18, 51 (1963).

27. S. S. Breese, R. Trautman, and H. L. Bachrach, Science, 150, 1303 (1965).

28. S. Sourirajan, Nature, 203, 1348 (1964).

29. W. J. Elford, Proc. Roy. Soc. (London), B106, 216 (1930); B112, 384 (1933); J. Pathol. Bacteriol., 34, 505 (1931); Trans. Faraday Soc., 33, 1094 (1937).

30. P. Grabar, L'Ultrafiltration Fractionnée, Hermann, Paris, 1943.

31. T. Svedberg and K. O. Pedersen, The Ultracentrifuge, Clarendon, Oxford, 1940; Johnson Reprint Co., New York, 1959.

32. H. Determann, Gel Chromatography, Springer, New York, 1968.

33. A. L. Shapiro, E. Viñuela, and J. V. Maizel, Biochem. Biophys. Res. Commun., 28, 815 (1967).

34. C. J. van Oss, Rec. Trav. Chim. Pays-Bas, 77, 479 (1958).

35. C. J. van Oss, H. Simonnet, and D. Annicolas, ibid., 78, 425 (1959); Compt. Rend., 248, 460 (1959).

36. S. Katz and I. M. Klotz, Arch. Biochem. Biophys., 44, 351 (1953).

37. G. A. J. van Oss and J. H. M. Koopman van Eupen, Rec. Trav. Chim., 76, 390 (1957).

38. S. Salminen, Dissertation, Helsinki, 1961; Ann. Med. Exptl. Biol. Fenniae, 39, Suppl. 4 (1961).

39. W. F. Blatt, S. M. Robinson, and H. J. Bixler, Anal. Biochem., 26, 151 (1968).

40. H. Paulus, ibid., 32, 91 (1969).

41. S. Sourirajan, Ind. Eng. Chem. Process Design Develop., 6, 154 (1967); 7, 548 (1968).

42. G. Farese and M. Mayer, Clin. Chem., 16, 280 (1970).

43. M. Stoliaroff, C. J. van Oss, G. Martin, and J. G. Gauduchon, Rev. Immunol., 22, 5 (1959).

44.   J. P. Agrawal and S. Sourirajan, Ind. Eng. Chem., 61, 62 (1969).

45.   A. Gouveia and K. A. H. Hooton, Chem. Eng. Progr. Symp. Ser.,
      64 (90), 281 (1968).

46.   K. C. Channabasappa, ibid., 65 (97), 140 (1969).

47.   B. J. Weissman, C. V. Smith, and R. W. Okey, ibid., 64 (90), 285
      (1968).

48.   Ultrafiltration for Laboratory and Clinical Purposes, Amicon Corp.
      Publ. No. 403, 1970, p. 8.

49.   J. P. Thiéry, C. J. van Oss, L. Salomon, and M. P. Doucet, Compt.
      Rend., 239, 1010 (1954); 239, 1096 (1954).

50.   D. O. Cliver, Appl. Microbiol., 13, 1 (1965).

51.   C. J. van Oss, A. Scheinman, and J. E. Lord, Nature, 215, 639
      (1967).

52.   C. J. van Oss and P. M. Bronson, Anal. Biochem., in press.

53.   J. W. McBain, Colloid Science, Health and Co., Boston, 1950,
      pp. 121-130.

54.   B. Erschler, Kolloid Z., 68, 289 (1931).

55.   S. Trautman, Biol. Méd., 37 (11,12) (1948).

56.   L. Ambard and S. Trautman, Ultrafiltration, Charles C. Thomas,
      Springfield, Ill., 1960.

57.   C. J. van Oss and N. R. Beyrard, J. Chim. Phys., 60, 451 (1963).

58.   C. J. van Oss, Dutch Pat. 74,531 (1954); British Pat. 721,087 (1954);
      Can. Pat. 548,246 (1957); Dutch Pat. 79,743 (1955).

59.   T. K. Sherwood, P. L. T. Brian, R. E. Fisher, and L. Dresner,
      Ind. Eng. Chem. Fund., 4, 113 (1965).

60.   P. L. T. Brian, in Desalination by Reverse Osmosis (U. Merten, ed.),
      M. I. T. Press, Cambridge, Mass., 1966, p. 161.

61.   S. Kimura and S. Sourirajan, Ind. Eng. Chem. Process Design
      Develop., 7, 41, 539, 548 (1968).

62.   S. Loeb, Desalination, 1, 35 (1965).

63.   D. G. Thomas and J. S. Watson, Ind. Eng. Chem. Process Design
      Develop., 7, 397 (1968).

64.   A. S. Michaels, U. S. Pat. 3,173,867 (1965).

65.   E. H. Sieveka, in Desalination by Reverse Osmosis (U. Merten, ed.),
      M. I. T. Press, Cambridge, Mass., 1966, p. 239.

Chapter IV

BIMOLECULAR LIPID MEMBRANES

H. Ti Tien

Department of Biophysics
Michigan State University
East Lansing, Michigan

and

Robert E. Howard

Department of Pathology
Medical School at San Antonio
The University of Texas
San Antonio, Texas

I.   INTRODUCTION

The principal aim of this chapter is to present a practical account of experimental arrangements and procedures currently being used for the study
of black lipid membranes of about bimolecular thickness in aqueous solution.
Therefore, the chapter is not written as a review but rather as an introduction to the subject for interested investigators who may wish to use these
ultrathin (~ 100 Å) membranes in their own work. The materials presented
in this chapter have been derived mainly from the published papers of many
investigators and ourselves. In particular, we wish to mention the work
initially carried out by Rudin, Mueller, Tien, and Wescott in Philadelphia,
and later in the laboratories of Haydon, Thompson, and Tien. For the
reader who desires a more general account of the work on these black lipid
membranes, there are a number of reviews available [1-4].

The limiting structure of black lipid membranes (BLM) (also known as
bimolecular or bilayer lipid membranes) has frequently been depicted as

very similar to the bimolecular leaflet first suggested by Gorter and Grendel for the structure of the red blood cell membrane [5]. Therefore, it seems appropriate to give a brief summary of the historical developments leading to the bimolecular leaflet concept and its relationship to general membrane phenomena.

Overton found that lipid-like substances readily diffused across cellular boundaries, and suggested in 1895 that the cellular membrane might contain lipids [6]. In the 1920's, a classic study by Gorter and Grendel demonstrated sufficient lipid in red blood cells to form a monolayer film with twice the surface area of the extracted cell ghosts [5]. Interfacial tension, studied in unicellular marine organisms, was interpreted as requiring the presence of protein as well as lipid. These and other considerations led several investigators to postulate in the 1930's that the basic chemical structure of the plasma membrane is a bilayer lipid leaflet with protein adsorbed to it in some fashion [7]. This general concept of membrane structure, refined and modified by Robertson and others [8,9] on the basis of additional experimental data, is now widely accepted if not definitely proved.

An understanding of biological membranes in physicochemical terms requires assessment of the contributions of specific molecules and molecular aggregates to various functional properties and structural details. Such correlations are made difficult by the varied ultrastructural relations, multiple functions, and complex chemical composition of natural membranes, but would be facilitated by an adequate model system.

Many different systems have been studied and extensively proposed as models of biological membranes. Beutner used polar oils to form membranes in the first such study at the beginning of this century [10]. Collodion films were used by Michaelis and Weech [11], cellophane was used by Teorell [12], and synthetic ion exchange resin sheets were used by Sollner [13] and others to study membrane characteristics. Tobias et al. studied a model membrane formed by supporting lipid on a Millipore filter [14]. More recently, Bangham et al. have studied microvesicles of bimolecular thickness [15,16].

Many other interfacial systems have been studied and have contributed to our understanding of membrane phenomena. These include interfaces between aqueous and lipid solutions [17,18]; micellar and colloidal suspensions [19,20]; and monolayers of lipid, protein, or synthetic polymers at the air-water interface [21,22]. The fundamental principles of orientation and interfacial behavior of amphiphilic molecules which were developed by Langmuir, Adam, Harkins, McBain, and Rideal through such studies have provided the basis for all subsequent progress in the study of membrane phenomena.

Although the construction and study of experimental models has proved scientifically useful, none of the above-mentioned models, with the possible exception to microvesicles, meet all of the criteria of analogy to natural

membranes which a fully adequate model membrane should possess. These
criteria may be listed as follows:

Chemical composition	(lipid, protein)
Dimensional requirement	(thickness about 100 Å or less)
Interfacial environment	(water-membrane-water)
Physical state	(stability and liquid-like)
Operational	(dissimilar aqueous phases separated; ease of manipulation and modification)

It is obvious that the chemical composition of any adequate model should
be similar to the composition of biological membranes, but it is perhaps less
obvious that the model must also possess dimensions similar to those of its
natural counterpart, i.e., a thickness of about 100 Å. This has been the
least frequently met criterion in previous models. Biological membranes
generally exist as liquid structures bounded on either side by an aqueous
phase; thus there are two coexisting liquid-liquid interfaces (or a biface)
associated with each membranous structure. From the viewpoint of interface science, biological membranes and bimolecular lipid membranes may
be considered as the limiting case of a three-component liquid system:
aqueous phase-lipid phase-aqueous phase, in which the lipid phase has been
reduced to molecular dimensions. Biological membranes are remarkably
stable structures, maintaining their integrity for many hours or even days,
even when separated from their parent cell. Similarly, an adequate model
should possess sufficient stability to enable its study. Characteristically,
biological membranes separate dissimilar aqueous phases, e. g., the cyto-
plasm and extracellular fluid are separated by the plasma membrane. Frequent structural asymmetry of the two sides of the membrane has been shown
by electron microscopy. Also, many biological membranes function in a
vectorial or directional manner, e. g., glucose accumulation in red blood
cells by carrier-facilitated or active transport. These three characteristics
impose corresponding requirements on any model membrane system: The
model should be capable of separating dissimilar aqueous phases, it should enable the formation of structural asymmetry, and it should provide conditions
such that unidirectional vectorial functions, such as transport, are assayable.

Bimolecular lipid membranes in aqueous media theoretically fulfill the
criteria of analogy to biological membranes outlined above and should facilitate both the study of the physicochemical properties of the membrane itself
and of transmembrane processes such as permeability, electrical transient
phenomena, and energy transduction.

The approach on which essentially all subsequent work has been based
was first reported by Mueller, Rudin, Tien, and Wescott in 1961 [23].
These investigators described the formation of a membranous structure with
aqueous interfaces formed from mixed bovine brain lipids. They showed by

optical, electrical, physical, and electron microscopic methods that these membranous structures approximated what would be expected for a lipid bilayer or a cell membrane [24]. A great amount of subsequent investigation has confirmed that these structures serve as a unique model system for elucidating the behavior and properties of biological membranes [1-4].

There are three specific objectives of this chapter. The first is to present various useful methods for the formation of bimolecular lipid membranes with aqueous interfaces --with sufficiently detailed information on apparatus, reagents, and technique that the investigator initiating studies on bimolecular lipid membranes will have a ready source of reference and may satisfactorily obtain results. The second aim is to provide selected and proven methods for the measurement of a wide variety of properties of the BLM system, with comments on the recognized methodological and technological pitfalls of each. The third aim is to stress the fact that black lipid membranes (or films) are a new type of interfacial adsorption phenomenon. Apart from the inherent interest of this new experimental structure from a biological viewpoint, it is also a useful tool for the understanding of the physics and chemistry of lipids and is relevant to the further development of interfacial chemistry and colloid science. In the past the study of BLM has been oriented mainly toward the elucidation of biological problems. It is hoped that this chapter will also stimulate some interest in additional physicochemical characterization of the BLM system.

II. FORMATION OF BIMOLECULAR LIPID MEMBRANES (BLM) WITH AQUEOUS INTERFACES

A. General Principles

The formation of a bimolecular lipid membrane (BLM) with aqueous interfaces is conceptually simple (Fig. 1): A hydrophobic support containing an aperture is immersed in an aqueous solution. A (thick) membrane of amphiphilic lipid is formed on the aperture and under favorable conditions spontaneously thins by draining centrifugally to the Plateau-Gibbs border (i.e., the lipid solution around the margin of the aperture). At equilibrium a black membrane of bimolecular thickness persists with an aqueous interface on either side of the membrane. Depending on the properties to be investigated, BLM can be formed either in a single aqueous compartment or separating two aqueous phases.

Amphiphilic lipid molecules are necessary in the bulk phase solution used to form the membrane. Amphiphilic lipids contain a water-soluble polar group and a nonpolar hydrocarbon portion with a very low affinity for water. When hydrated, a favored conformation of such amphiphilic lipids is a lamellar liquid crystalline phase ("neat" or smectic mesomorphic form) in which the lipid molecules are arranged in layers (lamellae) with their polar groups in a regular array, as in a crystal, while their hydrocarbon portions are in a liquid state. At the interface between bulk lipid phase and aqueous

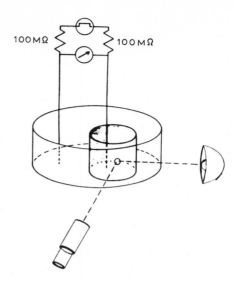

Fig. 1. The original setup used by initial workers for the formation of a bimolecular lipid membrane (BLM). The inner chamber is a 5-cm^3 polyethylene pH cup resting inside a glass petri dish (60 X 20 mm). Aqueous solution level is above the aperture; the BLM-forming solution is brushed on the aperture and observed at 10X magnification under reflected light. Conventional pulsing and recording circuit is also shown. (Reproduced by permission, from Recent Progress in Surface Science, Vol. 1, Academic, New York, 1964, p. 383. See Ref. [4].)

phase, the polar groups of the lipids face outward and contact the aqueous solution. In the limiting case of the BLM, only a single lamella continues to separate the aqueous phases on either side of the membrane (Fig. 2). The molecular configuration in the BLM is thus pictured as being very similar to the Gorter-Grendel concept for the structure of the plasma membrane [5]. Since BLM are ordinarily formed in aqueous solutions containing no dissolved proteins, there is no layer of adsorbed protein as was proposed later [7]. An arrangement of lipid molecules contrary to that shown in Fig. 2 is quite unlikely, as has been shown by an approximate thermodynamic argument based on principles of surface chemistry developed for adsorption at air-water and oil-water interfaces [25]. If protein molecules present in the aqueous phase are to be incorporated into the membrane, they must interact with the oriented amphiphilic lipid. In the older model [26], protein molecules were assumed to be bound electrostatically to the polar groups of the lipid. However, it has been suggested recently that in biological membranes proteins are bound mainly by hydrophobic interactions [27].

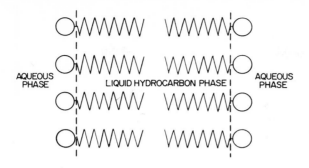

Fig. 2. Probable structure of a BLM in aqueous solution. Diagram
illustrates the orientation of lipid molecules containing polar groups at
membrane solution interfaces. Open circles and zig-zag lines represent,
respectively, polar and hydrophobic portions of the molecule. The interior
of the BLM is believed to be liquid-hydrocarbon-like.

B. Preparation of BLM-Forming Solutions and Other Reagents

1. Mixed Lipid Extracts

Solutions of mixed lipids extracted from biological materials were used
in the first successful formation of black lipid membranes and have been used
in many subsequent successful investigations. The initial investigators used
a mixture of brain lipids and proteolipids extracted with chloroform and
methanol from ox brain by a method modified after Folch and Lees [28].
After a repeated series of purification steps including centrifugation, filtration, emulsification, drying, and washing vs. an aqueous phase, the purified
lipid extract was dissolved in chloroform : methanol (2 : 1, v/v) for use as the
membrane-forming solution. A detailed procedure is given in Appendix A.
It may be mentioned that various hydrocarbons, silicone oils, or other substances can be utilized on an empirical basis to stabilize the membranes and
allow for more rapid thinning. Despite the fact that the potentially complex
composition of membranes formed from such mixed lipid extracts would not
permit exact correlation of physical characteristics with the chemical nature
of the membrane, the physical characteristics of BLM formed from such
mixed lipid extracts quickly verified their analogy to biological membranes.
Mixed lipids obtained from a variety of other sources can also be utilized.
The composition of some lipid solutions which have produced stable BLM is
given in Table I.

The phase partition method and its various modifications have been used
most frequently and most generally to obtain mixed lipid extracts. The modification by Howard and Burton [29] is rapid and has been successfully used
in preparing mixed lipid extracts from whole tissues or isolated membranous

TABLE I

Composition of Lipid Solutions Used in Bimolecular Membrane Formation[a]

Lipid	Solvent	Aqueous solution	Temp, °C	Remarks
Brain lipids + α-tocopherol (also with cholesterol)	$CHCL_3 : CH_3OH$ (2:1)	0.1 N NaCl or KCl	30–47	See Appendix A
Egg lecithin (also with cholesterol)	n-Decane	Various salt solutions	20–36	
Oxidized cholesterol	n-Octane	Various salt solutions	20–50	See Appendix B
Chloroplast extracts	n-Octane (60%) n-Butanol (40%)	Various salt solutions Various salt solutions	20–35	
7-Dehydrocholesterol	Various hydrocarbons	0.1 N NaCl	30	
Cholesterol (0.02 M) + dioctadecyl phosphite (DODP) (0.0027 M)	n-Dodecane	0.1 N NaCl	25	DODP from Hooker Chemical Co.
Cholesterol (0.026 M)	n-Dodecane	Various salt solutions +HDTAB[b]	25	HDTAB from Eastman Kodak Co.
Cholesterol (0.024 M) + dodecyl acid phosphate (DAP) (0.01 M)	n-Dodecane	Various salt solutions	25	DAP from Hooker Chemical Co.

[a] For further details, see Ref. [4].
[b] Hexadecyltrimethylammonium bromide.

fractions in a variety of species. Fresh tissue or tissue fraction is homoge-
nized for 2 min in 20 parts (v/w) of a nitrogen-saturated chloroform : methanol
mixture (2 : 1, v/v) in a conical glass homogenizer at $4°C$. The homogenate
is filtered through a Whatman No. 1 filter paper or a coarse sintered glass
funnel under a nitrogen atmosphere. Two volumes of distilled water are
added to the supernatant fluid, and the mixture is emulsified by shaking. The
lower organic phase, which contains a mixture of extracted lipids, is re-
covered after brief centrifugation to break the emulsion. In addition to being
rapid, this method requires fewer steps than the original method [24] and decreases the likelihood of lipid oxidation. The lower phase mixed lipid extract
may be used directly or it may be dried under reduced pressure under nitrogen and redissolved in 5.2 volumes (relative to the original tissue or tissue
fraction weight) of chloroform : methanol (2 : 1, v/v) solvent. The resulting
solutions are approximately 2% in mixed lipids.

A number of workers (see Table I) have experimented with various lipids
and other materials as well as solvents. In addition, large numbers of liquid
additives of various types have been tested for their ability to stabilize or
otherwise improve BLM formation. Hydrocarbons such as n-tetradecane;
α-tocopherol; silicone oils; fatty acids such as caprylic acid; fatty acid methyl
esters; and conjugated hydrocarbons such as squalene have been successfully
used as additives or as solvents in lieu of chloroform : methanol. It would appear that other nonaqueous solvents may be used alone or in combinations
provided that they have a sufficiently small aqueous : nonaqueous solvent distribution coefficient, that they be liquids with sufficiently low viscosity, and
that they solubilize sufficient lipid (usually about 0.5-2%).

In addition to the formation of BLM from solutions containing phospholipids, an "oxidized cholesterol" solution is known to produce very stable
black films over a wide temperature range [30]. The oxidized cholesterol
solution has been preferred by many investigators because of its simplicity
of preparation. The procedure for preparing the oxidized cholesterol BLM-
forming solution is given in Appendix B. This particular BLM-forming solution is recommended to those interested readers initiating study in the area
of black lipid membranes. For those interested in investigating photoelectric
effects in BLM, the scheme for extraction of lipids and pigments from fresh
spinach leaves and the preparation of a membrane-forming solution are given
in Appendix C.

2. Purified Substances

The desirability of using a chemically characterized, two-component
lipid solution which would better permit correlation of physical characteristics with the chemical properties of the BLM has been alluded to above. The
first successful generation of BLM from a two-component system was reported by Hanai et al. [31]. They used a highly purified egg phosphatidyl-
choline dissolved in an n-hydrocarbon solvent (Table I). Similar lipid-
hydrocarbon solutions have been used by the majority of subsequent workers.

All of the BLM-forming solutions prepared with mixed lipid extracts, and most of the solutions prepared with purified substances, have contained phospholipids. However, it should be noted that BLM can be generated from a variety of compounds other than phospholipids, as indicated in Table I. Of importance particularly to the field of interfacial chemistry, it has been shown that stable BLM can be formed from a solution containing purified cholesterol together with pure synthetic surface-active agents, in a hydrocarbon solvent [32]. These findings demonstrate clearly that the formation of BLM is a general phenomenon which exhibits only a broad requirement for an appropriate amphiphilic compound. It may therefore be expected that a wide variety of pure, well characterized, readily available surfactants may be utilized in BLM systems.

3. Problems with BLM-Forming Solutions

Although a wide variety of BLM-forming solutions have been successfully used, any given solution, although prepared according to recipe, may not satisfactorily form BLM. Unsatisfactory BLM formation may result from failure to form the initial thick membrane, failure to thin, instability and early rupture of the membrane either during thinning or immediately after BLM formation, or from gross inconsistencies in the properties of membranes formed from the same or similar solutions.

Efforts to prepare BLM-forming solutions whose compositions duplicate those given by other workers, e. g., compositions as in Table I, have frequently been frustrated due to trace impurities in either lipid or solvent. Although no quantitative data are available, this seems particularly true when using biological phospholipids, complex additives, and synthetic surfactants which have been oxidized, excessively aged, or insufficiently purified. For this reason careful purification, redistillation, etc., are generally advisable, although it should be noted that the use of commercially available lipids and solvents without further purification has at times been successful [32]. BLM-forming solutions prepared from mixed lipid extracts in a chloroform : methanol solvent system tend to deteriorate rapidly unless refrigerated and maintained under nitrogen. Conversely, BLM-forming solutions prepared from highly purified phospholipids in a hydrocarbon solvent, with or without the addition of cholesterol, may remain usable for over a year after storage under nitrogen at $4°C$ in sealed ampules.

Efforts to establish the optimum composition of a new BLM-forming solution have been relatively simple in our experience, although frequently somewhat tedious. There is no method presently available which can predict the necessary chemical composition for a BLM-forming solution which will successfully form stable BLM. There has been a paucity of theoretical work on BLM stability. The possible criteria for BLM stability include: (1) low BLM interfacial tension, probably within a range of 0-6 dyn/cm and, (2) a specific distribution of space charge of molecules.

After the mixed lipid extract or individual amphiphilic compound has been obtained and purified, it is necessary to select an appropriate solvent.

The normal aliphatic hydrocarbons (C_6 - C_{16}) have proved most generally useful as solvents. These hydrocarbons are all liquid at room temperature and are very poorly soluble in water. Although n-octane, n-decane, and n-dodecane have been used most frequently, the choice of hydrocarbon length does not appear to be critical. Chloroform : methanol in 2 : 1 and 3 : 2 volume ratios have previously been used as solvents but are somewhat unsatisfactory unless an additional liquid additive is used. Other substances which have been used include cyclohexane, carbon tetrachloride, benzene, toluene, and halogenated hydrocarbons.

Final concentration of lipid in successful BLM-forming solutions is generally at or near saturation, that is, ca. 0.5-2%. The exceptions are those BLM-forming solutions in which the lipid solvent is highly water-soluble, e. g., chloroform : methanol. It has been suggested that the inability of dilute lipid solutions to form stable BLM may be due to a decreased interfacial pressure which is insufficient to compress the lipid molecules at the BLM interface [32].

4. Aqueous Phases

BLM have been successfully formed in a wide variety of aqueous solutions, ranging from distilled water to buffered multicomponent solutions simulating physiological fluids [36]. Table II gives a summary of the composition of some aqueous solutions reported in the literature. The aqueous phase most frequently used has been an unbuffered solution of a simple electrolyte in distilled water, e. g., NaCl. It is obvious that BLM may be formed in a great variety of aqueous solutions.

The major aspects of aqueous phase composition are the molar concentration (molarity) of the various constituent substances, the pH, and (when salts are present) the ionic strength. Because osmotic pressure gradients can cause BLM deformation and volume flow of water, osmolalities should generally be the same in each aqueous phase when the BLM separates two aqueous compartments.

Although the composition of aqueous phases may vary widely, the normal precautions of interface chemistry are necessary to prevent trace impurities of surface-active agents, trace contamination with metals and ions, and lipid oxidation. The use of redistilled water, carefully cleaned glassware, analytical reagent grade salts, and carefully purified solutes will generally suffice. The pH of the aqueous phase is generally not critical, and stable BLM may be formed over a wide range of pH, as indicated in Table II.

The quantity of aqueous phase is in all cases very much greater than the quantity of BLM-forming solution. Because of this great difference, the small but finite solubility of lipids and solvents in the aqueous phase may lead to rapid alteration of the membrane with subsequent instability. Stability of BLM may be improved by equilibrating the aqueous phase with the lipids and hydrocarbons to be used in the BLM-forming solution. Small gas bubbles derived from dissolved gases in the aqueous phase may be prevented

TABLE II

Composition of Aqueous Media

Aqueous media	Components	Concentration, mM/liter
Histidine–buffered Tyrode solution	NaCl	129
	NaHCO$_3$	6
	Na$_3$PO$_4$	2
	KCl	4.2
	CaCl$_2$	5
Adjust to pH 7.3 at 37°C	MgCl$_2$	1.6
with HCl or NaOH	Histidine	5
Tris-buffered saline	NaCl	100
	Tris (hydroxymethyl) aminomethane	50
Adjust to pH 7.4 at 25°C	Mercaptoethanol (when present)	1
with HCl or NaOH	Dithiothreitol (when present)	1

by degassing the solution. When biological lipids or lipid extracts are used for BLM formation, lipid oxidation may be retarded by the addition of mercaptoethanol, dithiothreitol (2,3-dihydroxy-1,4-dithiobutane) at 0.001 M concentrations, or α-tocopherol.

The species or ionic strength of salts in the aqueous phase generally has no effect on the stability of phospholipid membranes. For example, Hanai et al. [31] formed stable BLM in salt solutions over a range of 1 X 10^{-4} to 4 N. Although BLM stability was not altered, Howard and Burton [29] have observed that in certain cases the membranes appeared more rigid with increased divalent cation concentration. In contrast, cholesterol-HDTAB (hexadecyltrimethyl ammonium bromide) membranes [33] showed marked sensitivity to changes in ionic strength. At ionic strength of 0.1 N, BLM could not be formed or would last only a few seconds. However, BLM were stable for 20 min or longer in aqueous phases of ionic strength 10^{-3} N or less.

C. Apparatus and Procedures for Forming BLM

1. BLM in a Single Aqueous Compartment

Several types of BLM investigations may best be made on a BLM formed with a loop or frame which is immersed in a single aqueous compartment. Such investigations include study of formation characteristics, stability, electron microscopic appearance, optical properties, and interaction of the membrane with macromolecular species or subcellular components. These methods involve generally easier BLM formation, less expensive and intricate apparatus, and fewer mechanical and technical problems. In addition, they obviate many of the problems of BLM separating two aqueous compartments which result from transmembrane gradients of hydrostatic pressure, osmotic pressure, chemical composition, or electrochemical force. For classification purposes these methods are named in operational terms, such as "dipping method," "painting method," "injection method," etc.

The "dipping" method is the most simple and has been used to form membranes for simple visual observation, photography, electron microscopy, and for testing the efficacy of membrane-forming solutions. Membranes are formed by dipping a thin metal, glass, polyethylene, or hair loop with a loop diameter of 1-2 mm into the lipid solution, and then transferring the membrane through air into a vessel containing an aqueous medium, such as 0.1 M NaCl at 30-50° C. Small Teflon O-rings may also be used for this purpose. The dipping method was used by early workers [34] to form BLM on a hair loop for electron microscopic studies.

The "painting" or "brush" method originally described [34] generally provides a more satisfactory method of membrane formation. The loop or O-ring is positioned within the aqueous phase and the membrane-forming solution is applied to it with a brush or other device. The initial investigators used a fine No. 3 sable hair artist's bright (brush) which was trimmed to two or three hairs' thickness, thus preventing application of excess BLM-forming solution. The brush is simply dipped in the appropriate membrane-forming solution, and a thin layer of lipid solution is spread over the aperture in the loop or O-ring.

Howard and Burton [29] have reported an improved painting method in which sable hair brushes are replaced by polyethylene or Teflon spatulas. Polyethylene stirring devices called "Plumpers" (California Corporation for Biochemical Research) are heated, and the softened broad end is spread and thinned by squeezing with smooth-jawed metal pliers to form a tapered blade. The end and sides of the blade are further shaped and trimmed with a razor blade. Similar spatulas can be formed from sheet Teflon with a razor blade. The plastic spatulas may be cleaned thoroughly after each use, they do not contaminate membrane-forming solutions, and they have a long useful life.

Another modification of the painting method was reported in which a short length of Teflon tubing was connected at the end of a small syringe, with the Teflon tubing being used as a brush. This method has the advantage that the BLM-forming solution may be contained within the syringe, and the disadvantage that it is satisfactory only with small apertures [33].

The "injection" method using an apparatus shown in Fig. 3 is perhaps the best available method for studying BLM in a single aqueous compartment [25]. The apparatus consists basically of two rectangular glass chambers arranged one inside the other. Within the inner chamber a mirror-finished Teflon O-ring (available from Chicago Gasket Company, 1271 North Avenue, Chicago, Ill.) is supported by platinum wires at the end of the glass capillary tubing and is used as a supporting aperture for the BLM. The rectangular glass tubes are culture tubes, the sides of which are optically flat. The end of the capillary tube is blown out to approximately the same diameter as the O-ring (4 mm), with the glass capillary mounted coaxially in the 12/30 ground joint. The female portion of the 12/30 ground joint is mounted coaxially with the long axis of the two rectangular chambers, so that the ground joint allows orientation of the BLM relative to the planar sides of the rectangular chambers. Glass tubing blown onto the culture tube permits introduction of fresh aqueous phase into the chamber, and a glass tube blown onto the capillary holder permits removal of waste aqueous phase from the chamber. The space between the two glass chambers serves as a constant temperature bath through which thermostated water may be circulated. A thin polyethylene tubing (PE-50, Clay Adams Company, 141 East 25th Street, New York, N. Y.) is attached to the glass capillary tubing at one end and to a three-way valve at the other. This valve is attached in turn to a syringe filled with BLM-forming solution and mounted on an infusion-withdrawal pump. An additional capillary for the injection of other test compounds is provided at the bottom of the system. Stray light reflections are minimized by painting the whole assembly black with the exception of one porthole for light paths. The membrane is illuminated by light from a microscope illuminator which is passed through an iris diaphragm, condenser, interference filter, and focusing lens. The membrane may be viewed by any low power, 10-40X, monocular, binocular, or stereo microscope, preferably with wide field oculars.

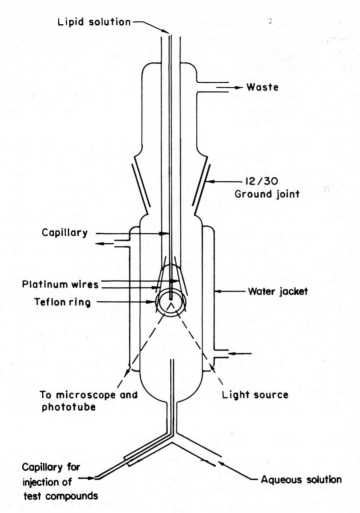

Fig. 3. Apparatus used for the investigation of formation characteristics and optical properties of bimolecular lipid membranes. This apparatus is also useful for carrying out stability and lipid-protein interaction studies (see text for details).

This apparatus is useful for the examination of large numbers of individual compounds or for mixtures of compounds prior to studies of BLM separating two aqueous compartments. The apparatus permits repeated application of membrane-forming solution to the aperture and rapid renewal of fresh aqueous phase under reproducible conditions. A modification of this method which may be called the "interface application" method has also been developed for preparation of BLM with relatively large areas [25]. The

method basically uses the chamber described above, which is initially filled
with aqueous phase to a level just below the Teflon O-ring. A few drops of
membrane-forming solution are introduced on the surface of the aqueous
phase, and the lipid phase-aqueous phase interface is raised above the level
of the O-ring by admitting more aqueous phase from below. BLM have successfully been formed on Teflon O-rings up to 1 cm in diameter with this
method. The method appears potentially capable of forming BLM with areas
of several square centimeters.

A "spinning" method has been suggested as useful for producing large
sheets of BLM [24]. In this method the membrane-forming solution is fed
into an enlarging frame surrounded by aqueous phase. With proper timing,
only the relatively strong BLM itself is "spun" into the frame. The mechanical strength of BLM could possible be improved by addition of mucopoly-
saccharides, proteins, etc., to the aqueous phase as stabilizing agents [24].

A "marginal suction" method for BLM formation similar to the method
reported for soap films has been reported [36]. The BLM is supported on a
Teflon O-ring (7 mm o.d. X 3 mm i.d.) which is drilled through from the
outside to the lumen side and supported by the needle of a commercial microliter syringe (Hamilton No. 1705). The needle and O-ring are immersed in
thermostated aqueous medium after the syringe has been filled with BLM-
forming solution. Membranes are formed by slowly injecting solution from
the syringe until the O-ring lumen is completely filled with a drop of BLM-
forming solution. The bulk phase lipid solution is then withdrawn slowly back
into the syringe, causing the drop of BLM-forming solution to become a
biconcave disk. As the solution is further withdrawn by suction on the margin
of the disk, the concave surfaces meet and form a planar membrane, supported by an annulus of bulk phase lipid (Plateau-Gibbs border) as in other
methods.

2. BLM Separating Two Aqueous Compartments

Measurement of the electrical properties, permeability properties, and
bifacial tension of BLM require that the membrane separate two aqueous
compartments. Numerous types of apparatus have been described for this
purpose, but details will be given here for only four designs of apparatus. In
general, BLM separating aqueous compartments may be formed by methods
similar to those described in the last section. The only unique method introduced in this section is the "bubble-blowing" method for formation of spherical BLM.

The apparatus described by the initial workers [24] is shown in Fig. 1.
One aqueous compartment was a 5-ml polyethylene pH cup (Beckman). The
other aqueous compartment was a glass petri dish, 20 mm high by 60 mm in
diameter. The BLM was formed on a smooth circular hole in the wall of the
pH cup. The wall of the pH cup was thinned with smooth-faced heated pliers
to about 0.2 mm thickness, and the hole was produced by a heated needle
held in a drill chuck. The pH cup was held in place in the petri dish by means

of a spring clip.   The temperature of the solution in both chambers was regulated with a heating lamp.

The brush method was used to form BLM in this simple apparatus.   Both aqueous compartments were filled with aqueous phase, and a lipid film was spread from BLM-forming solution with a sable hair brush.   The early investigators performed most of the initial characterization of BLM with this simple apparatus, and it has been widely used by other workers [31,35].

The apparatus used by Hanai et al. [31] consisted of an inner compartment formed from a solid Teflon rod, with a square glass vessel as an outer compartment.   Howard [36] has described in detail the designs of two basic kinds of apparatus useful in a variety of BLM studies.   One is a cylindrical Lucite chamber especially constructed to permit optical investigation of the membrane, and the second is a small, interchangeable Lucite chamber unit designed to be used multiply with a thermostat block for the study of several different membranes, either simultaneously or in rapid succession.   These interchangeable chamber units were designed to be held in a thermostat block, with assemblies of electrodes and provisions for perfusion or sampling of aqueous phase placed in the chambers as required.   Methods are described which enable simultaneous mechanical, electrical, optical, and chemical operations and studies to be performed on the same membrane with this apparatus.   The reader is referred to Howard [36] for details.   A similar style of Lucite apparatus has been described by Huang and Thompson [37] in studies on water permeability of BLM.   A rather more elaborate version of their system, using Lucite components for aqueous compartments with a polyethylene partition with aperture for BLM support, has been described by Andreoli et al. [38].

Cass and Finkelstein [39] used the brush method with a somewhat different apparatus for study of water permeability of BLM.   One aqueous compartment of their apparatus consisted of a microliter syringe (Hamilton No. 7101-N), the needle of which had been bent at a 90° angle, and tipped with short concentric tubes of polyethylene tubing (PE-50/90/160).   The BLM was formed with the brush method over the freshly cut end of the polyethylene tubing, which was supported in an outer compartment consisting of a 100-ml clear Lucite box.

An apparatus useful for study of electrical properties in interfacial tension of BLM, utilizing the "injection" method of BLM formation, has been described.   The apparatus, shown in Fig. 4, incorporates the unique feature of using readily available Teflon sleeves (designed for ground-glass joints) to form and separate the two aqueous compartments, with the rest of the apparatus made of glass.   The front portion of the glass compartment is flattened to facilitate visual observation of the membrane.   The Teflon sleeve is held in place between the male and female portions of the ground-glass joint, and the aperture (about 1.6 mm in diameter) is punched in the Teflon sleeve, rather than being drilled [32].   Temperature control in this apparatus is achieved by flowing thermostated water through a glass coil in the outer compartment.

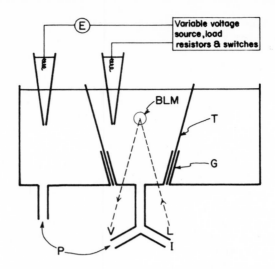

Fig. 4. Experimental arrangement for measuring the electrical properties and interfacial tension of bimolecular (black) lipid membranes. L–light
source, V–viewing tube, G–ground joints, T–Teflon sleeve, P–pressure
transducer connections, I–infusion-withdrawal pump.

The injection method [32, 33], as used on apertures in Teflon supports,
is more reproducible than the brush method. A 2- to 4-μl drop of BLM-
forming solution is injected onto the aperture by means of a 100-μl syringe
with a repeating dispenser attachment (Model PB-600-1, Hamilton, Whittier,
Calif.). A short length of polyethylene or Teflon tubing (1 mm^2 area at end)
is attached to the syringe and positioned in front of the aperture. With this
micrometer-type attachment a precise but minute volume of BLM-forming
solution may be repeatedly injected to form the BLM. Although membrane
formation with the injection method is basically a simple process, practice
and experience are necessary, as with the brush technique.

The "interface application" method has been described by Vandenberg
[40] for BLM separating two aqueous compartments. The apparatus
described consists of a Teflon sheet, containing the BLM-supporting aperture, which fits diagonally into a cuvette with square cross section. The
walls of the cuvette are siliconized and the edges of the Teflon sheet are
sealed against the cuvette with a dense hydrophobic material such as chloroform. The BLM is formed by lowering the aperture from an upper phase of
membrane-forming solution (e. g., lecithin-heptane) into the lower aqueous
phase, permitting a small portion of the aperture to remain in the membrane-
forming solution. As the aperture is lowered through the interface between
the membrane-forming solution and the aqueous phase, the interface is
simultaneously applied over both sides of the aperture to form the initial
thick film. This method is described as providing electrical insulation of

the two aqueous faces, and BLM with areas up to 50 mm^2. Contamination at the interfaces of aqueous phase with insulating oil, or aqueous phase with membrane-forming solution, causes marked limitation of the usefulness of this method, especially if precipitation occurs at the interface.

The "bubble-blowing" method permits formation of a spherical BLM which separates one aqueous compartment contained on the inside of the sphere from a second, external aqueous compartment. This method has been mentioned by a number of investigators [24, 32] and has been described recently in more detail [41]. The method, in its most simple form, consists of attaching a hypodermic needle (19-22 gauge, occasionally blunt-ended) to a 1-ml tuberculin syringe filled with warm aqueous phase, dipping the needle into membrane-forming solution, placing the needle in a vessel containing warmed aqueous phase, and "blowing a bubble," in which the first aqueous phase is separated from the second by a thin film of lipid derived from the membrane-forming solution. In one method described, a syringe microburet and a small-diameter polyethylene tube replace the tuberculin syringe needle, and the second aqueous phase in the vessel is formed as a density gradient. With this method a suspended or free-floating BLM sphere is formed, with a lens of bulk phase lipid covering one portion (usually the upper portion) of the sphere. Spheres of up to approximately 1 cm in diameter may be obtained with this method, but the spherical BLM offer no particular experimental advantage over planar BLM for optical, electrical, or interfacial tension studies.

Recently Tsofina et al. [42] described an apparatus for producing BLM which is similar to the chambers used by Deryagin et al. [43] and by Sheludko in thin soap film studies [44]. The apparatus consisted of an open chamber filled with lipid solution, and two devices for extruding spheroidal droplets of aqueous solution toward each other. One device was a length of tubing inserted through an opening in the bottom of the chamber. The other device, resembling a small funnel with a closed top and a side-arm outlet tube, was lowered into the chamber to a level just above the tubing. When spheroidal droplets of aqueous solution were simultaneously pushed part way out of each device, a lipid membrane remained at the area where the drops came together.

A combination of the monolayer technique and the dipping method, described by Takagi et al. [45], is illustrated in Fig. 5. A partition made from a sheet of Teflon (0.05 mm thick) with a hole (1 mm) at the center is placed between two halves of a trough, with the hole above the aqueous phase. The trough is first filled with an aqueous solution. After the surface has been cleared, a monolayer of a suitable lipid(s) is spread on the surface in the usual manner [22]. The Teflon partition is then lowered slowly into the aqueous solution. A BLM is thus formed over the hole, as shown schematically in Fig. 5(C). Precautions should be taken to keep the surface pressure of the monolayer constant in order to produce a stable BLM. This may be accomplished by adjusting the position of the barriers during the dipping operation.

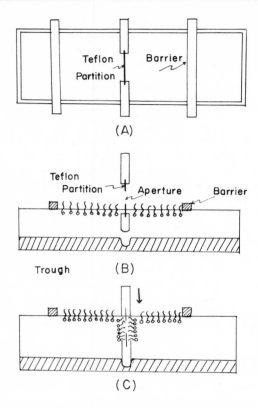

Fig. 5. Schematic views of an apparatus for the formation of BLM.
(A) Top view of a Langmuir trough with a Teflon partition at the center. (B)
Before dipping, a monolayer is formed at the air–water interface in the
usual manner. (C) The Teflon partition is lowered through the interface,
thereby forming the BLM.

D. General Considerations and Precautions

1. Theoretical Variables in BLM Systems

In any experimental system it is necessary to control or quantitatively
measure all variables in order to derive valid and meaningful information
from the system. The brief discussion of theoretical variables which is
presented in the following paragraphs is intended to serve as an introduction
for the interested reader new to the BLM field and is a brief review for
those already familiar with BLM apparatus and methods.

Some variables may be readily controlled, such as the aqueous media,
the gaseous phase (when it is present or important), the environmental
pressure, and the temperature. When a BLM is formed in a single aqueous

compartment, the various transmembrane gradients are generally nulled. When a BLM separates two aqueous compartments, transmembrane gradients may occasionally be nulled but must often be controlled or measured. The membrane itself, and the Plateau-Gibbs (P-G) border which surrounds and supports it, are variables which cannot be so well controlled. In a complete BLM system, the total quantity of lipid present would ideally be distributed between the P-G border and the BLM. However, the quantity of lipid present in the P-G border which surrounds and directly supports the membrane is vastly greater than the quantity of lipid in the membrane itself. Portions of the total lipid in the system may also be present as a thin film on the partition, as droplets or a film on the surface of the aqueous medium, in free solution, in micellar solution, or as a fine suspension within the aqueous medium. "Dipping," "brush," and "interface application" methods of BLM formation do not permit control of the amount of lipid initially introduced into the system. In contrast, the "injection," "bubble-blowing," and "marginal suction" methods enable measurement, with a microsyringe, of the total amount of BLM-forming solution introduced into the system.

The chemical composition of the P-G border may theoretically be controlled in the case of a membrane-forming solution containing only one component. However, successful BLM formation has thus far been demonstrated only with multicomponent solutions. This creates the problem of the relative distribution of a given component between the P-G border and the BLM. The problem is further compounded for substances with significant solubility in the aqueous medium. Such solubility introduces the possibility of a gradual but significant change both in P-G border composition and in the total quantity of material present in the P-G border. Conversely, material from the aqueous medium may become incorporated into the P-G border.

The aqueous medium in each compartment must be controlled for quantity (volume), composition, and homogeneity. The initial volume may be altered by gradual changes caused by hydrostatic or osmotic gradients, which may result in membrane deformation or volume flow of water or both. The osmolalities of each aqueous phase should therefore generally be equal. In most BLM systems, the quantity of aqueous medium is many orders of magnitude greater than the quantity of membrane-forming solution. Thus, substances present in trace quantities in the aqueous medium may nonetheless be present in quantities which are large relative to the quantity of lipid in the BLM.

The composition of the aqueous medium can be controlled prior to formation of the complete BLM system, but may be subject to change after exposure to membrane-forming solution. The major aspects of composition are the molar concentration of the various constituent substances, the pH, and, when ionizable substances are present, the ionic strength. Alteration in the composition of the aqueous medium may occur because of exchange of substances with the membrane-forming phase, by exchange of substances across the membrane to or from the other aqueous compartment, by electrochemical reaction, and (if the aqueous medium is exposed to the environment) by evaporation and loss of water or volatile solutes.

The homogeneity of the aqueous medium is of considerable importance in BLM systems, as homogeneity is assumed in many of the theoretical treatments of similar systems. It is also important operationally, since in determinations of BLM permeability the composition of the aqueous medium in the entire compartment is assumed to represent the composition at the aqueous-BLM interface. When two aqueous media of different composition are separated by a BLM, the movements of solutes and solvent along the respective concentration gradients will cause local inhomogeneities in the aqueous medium. In addition, the BLM itself may cause a local ordering of water, creating a boundary layer which will cause the true transmembrane concentration gradient to differ from that in the bulk phases. Experimental evidence shows that sufficiently vigorous stirring may overcome all but the last of these factors.

The usual BLM system contains a gaseous phase which is contiguous with the atmosphere over each aqueous compartment. This gaseous phase, however, may be part of a closed system or it may even be absent. When part of a closed system, its quantity, composition, and pressure must be considered. In the usual uncontrolled situation there is a virtually infinite quantity of gas phase at atmospheric pressure, composed of air or some superfusing gas such as nitrogen.

The BLM itself is the most important variable in the system and the most difficult to control. As they thin, newly formed lipid membranes (which are not BLM) contain a continually decreasing quantity of lipid. At equilibrium the quantity of BLM lipid can theoretically be estimated from the membrane volume by taking the product of membrane area and membrane thickness in a manner analogous to that used with the monolayer film balance. Optical and electrical methods are available which permit measurement of BLM area, and estimates of BLM thickness which are dependent upon assumptions discussed in a subsequent section. As an example, a BLM of 1.0 cm^2 size and 50 Å thickness would have a volume of 5×10^{-7} cm^3. If a reasonable value, e. g., 50 Å^2, is assumed for the surface area per lipid molecule, this same BLM would contain 4×10^{14} molecules or about 6.4×10^{-10} moles of lipid.

The composition of the BLM is of particular interest although not readily amenable to analysis. Because the quantity of lipid in the BLM is minute and because multicomponent solutions must presently be used to form BLM, the problems of composition which were previously discussed for the P-G border becomes greatly magnified. Because BLM area is large relative to its volume and thickness, the dissolution of substances from BLM into either aqueous medium is hastened. Likewise, the absorption of substances from the aqueous phase into the BLM is rapid. Because there is a very great difference between the volumes of the membrane and of the P-G border, a large change in the composition of the BLM due to altered distribution of components will result in only a minute change in the composition of the P-G border. If, as is currently believed, the BLM is in dynamic equilibrium with its supporting P-G border, gradients of compositional difference might exist between the center of the membrane and the P-G border. Studies made recently [46] on BLM formed from radioactive phosphatides, cholesterol, and hydrocarbons indicate that the molar concentration of liquid hydrocarbon filler may be severalfold greater than that of the phospholipid or cholesterol membrane constituents.

Thin lipid membranes in aqueous media, including BLM, contain two interfaces between the membrane-forming lipid phase and the aqueous media. The term "biface" has been used to describe and to emphasize the coexistence of these two interfaces. The biface is generally considered to be symmetrical in composition, but it is theoretically possible for one interface to be of different composition than the other, separated by a hydrophobic region of molecular thickness (~ 50 Å).

Ordinarily, the environmental pressure exerted on the BLM results from the atmospheric pressure of the continuous gaseous phase plus the hydrostatic pressure of the aqueous medium. It is theoretically possible, however, to alter the environmental pressure on the entire BLM system, although no such experiments have yet been reported.

It should be stressed that temperature is a particularly important variable in BLM systems, because the various transitional forms of liquid crystalline phase which form BLM are usually quite temperature dependent. The formation, thinning, and stability of BLM are markedly affected by small temperature changes, as discussed in other sections. Such temperature changes can also induce significant alteration in solubility and diffusion rates of substances within the system, and in the viscosity of the liquid phases. The effect of temperature on such BLM properties as electrical resistance and water permeability will be discussed in subsequent sections.

Transmembrane gradients of molecular and ionic concentration, electromotive force, and hydrostatic pressure are variables which have received considerable experimental attention. The first two types of gradients are closely related, with a single substance (e. g., NaCl) often giving rise simultaneously to different gradients. The concentration gradient for any diffusable molecule or ion provides the driving force for its transmembrane flux and for the osmotic flux of water in the opposite direction. Concentration gradients of nondiffusable substances drive only the osmotic flux of water. Ionic concentration gradients may give rise to a transmembrane potential. Additional electrical potential may be derived from an externally applied transmembrane voltage. The effects of ionic or electric potential gradients are discussed in Secs. IV and V. Hydrostatic pressure gradients may also exist across the BLM and have been utilized in measurement of bifacial tension, as discussed in Sec. VI.

2. Operational Variables in BLM Systems

There are several general operational parameters to consider in the assembly of any BLM system. It should be easy to thoroughly clean the chamber components, aperture supports, thermostating provisions, electrodes, etc., in order to prevent chemical contamination. Systems should be designed so that they will be relatively easy to construct and modify. Although BLM systems may require many functional components, any component should be removable for repair, substitution, modification, or calibration without disrupting and disassembling the entire system. Disassembly and reassembly of the entire BLM system should be simple and straightforward procedures.

The optical, electrical, and sampling methods which have been used to measure membrane thickness and area, electrical properties, and permeability properties may impose further requirements on the design and assembly of a BLM system for any particular investigation.

There are two optical parameters which must generally be provided for in any BLM system: an optical path for visualization of the membrane, and provision of shielding to eliminate extraneous light. Visualization of the membrane requires an optical path such that the incident and reflected light rays lie in a plane which is perpendicular to the plane of the BLM. Shielding can be accomplished by painting the apparatus with a flat black, or by constructing the apparatus of black Lucite. Where possible, optical surfaces should be inclined to prevent spurious reflections, and stray or transmitted light from the light source should be trapped.

All BLM systems should be thermostated, preferably with provision for both heating and cooling. The measurement and/or regulation of temperature should ideally be achieved with the thermometer or thermistor placed directly in the aqueous medium, adjacent to the BLM. This is particularly true where large quantities of aqueous medium are being exchanged through a small aqueous compartment, and in conditions of changing thermostasis.

Accurate electrical measurements are necessary for determination of several experimental variables, such as the pH and ionic strength of each aqueous medium and the transmembrane electrochemical potential, resistance, capacity, and dielectric breakdown voltage. Such measurements require that no ion- or electron-conductive pathways exist between the two aqueous phases except via the BLM. This implies that the entire BLM chamber and BLM-supporting partition be of nonconducting material, and that all electrodes, valves, thermometers, alignment pins, stirring provisions, etc., be insulated from each other, from ground, and from their counterpart on the opposite side of the BLM. In addition, high sensitivity measurements require electromagnetic shielding of all conducting leads and electrodes from extraneous and uncontrolled electromagnetic fields, and from each other, by a common ground or a driven shield.

The materials which are usable in construction of BLM chambers and supports are markedly limited by the requisites of chemical inertness to both aqueous and organic solvents, electrical insulation, and optical transparency. Glass components, because they possess these requisites, are useful when they can be incorporated into the system. Cast Lucite (Plexiglas) is electrically nonconductive, transparent, and resistant to many organic solvents. Because Lucite may be easily machined and cemented, it has frequently been

used in lieu of glass. However, Lucite should not be used when halogenated
solvents such as chloroform are used as lipid solvents. Teflon, polyethylene,
and polypropylene have the requisite chemical and electrical properties and
may be machined or cast, but they lack transparency and rigidity. Teflon
has proved particularly useful in O-rings and in aperture-containing sheets
and cups for BLM support.

3. Procedural Problems and Precautions

Despite the apparent simplicity of forming BLM in aqueous media, several experimental problems may be encountered. Not only may these problems plague the beginner, but they may occur unexpectedly to puzzle the experienced BLM investigator. Although the existence of some of these problems has been denied by some investigators, it is our feeling that they deserve
some special comments. Some of these major factors are discussed below.

a. Inadequate cleaning of all apparatus is a frequent problem which may
impair or prevent BLM formation or decrease BLM stability. Traces of
surface-active agents may be derived in a BLM system from a variety of
sources, including plasticizers, curing agents, catalysts, and mold release
agents containing in plastic portions of a newly constructed system; body oils
or other contaminants from handling; and residual oxidized lipids or minute
traces of detergent due to incomplete washing and rinsing of the system components. Glass components may be adequately cleaned by overnight immersion in fresh sulfuric acid-dichromate cleaning solution, followed by repeated
rinsing in distilled water. Surface-active agents can be completely removed
from new Lucite by washing and exposing it to ultrasound and hot aqueous
detergent, rinsing it for several hours in hot running tap water, and then
boiling it in distilled water. Teflon, polyethylene, and polypropylene pieces
can be washed by extraction with chloroform : methanol (2 : 1, v/v) overnight
and boiling in distilled water. Once BLM systems components are initially
cleaned of surface-active agents, they can usually be maintained in an adequately clean state by brief washing with detergent, rinsing with hot tap water,
rinsing with distilled water, and drying in air prior to use. Precautions such
as routinely storing glass or Lucite components under distilled water or storing Teflon or polymer components under petroleum ether are generally
unnecessary. However, components exposed to the atmosphere in some
laboratories for longer than a day or two may become contaminated and require cleaning again before use.

b. Thin lipid membranes and BLM can generally be formed on clean
untreated loops, Teflon O-rings, or aperture-containing partitions. "Preconditioning" of the aperture may, however, improve membrane formation
and stability. Preconditioning is achieved by air-drying a small amount of
membrane-forming solution in the aperture before assembly of the system
and introduction of aqueous solution.

c. The lipids and hydrocarbons in BLM-forming solutions are generally
less dense than water, and quantities in excess of that required for the membrane and P-G border will thus generally move to the surface of the aqueous

medium. With the brush method, if too much BLM-forming solution is applied, the excess may flow upward and rupture newly formed membranes. This upward flow with rupture of the membrane is particularly a problem with untrimmed sable brushes and with apertures greater than 5 mm in diameter. Teflon or polyethylene spatulas, or the Teflon tubing-on-syringe, hold less lipid and generally do not create this problem with the brush method. A similar problem occurs with the injection method if too large an aperture is used or if too much BLM-forming solution is ejected. Conversely, too little BLM-forming solution in the brush or injection method will detract from ready membrane formation because a sufficient supporting P-G border and thick membrane cannot be formed. With the bubble-blowing method, too much lipid will result in an excessively large lens of bulk lipid solution attached generally at the upper surface of the spherical BLM.

d. Oxidation of lipids in the BLM-forming solutions by prolonged exposure to air may make the formation of membranes difficult or impossible, may lead to premature rupture of the thinning membrane, or may result in BLM of decreased stability. Freshly prepared phospholipid solutions using hydrocarbon solvent are generally stable for several months when stored under nitrogen in a freezer ($< -25^{\circ}$C) and maintain their membrane-forming capacity even if exposed to air at room temperature for a working day. BLM-forming solutions made with chloroform : methanol are generally less stable, becoming unusable after 4-6 weeks in a refrigerator, or after a few hours of exposure to air at room temperature. In contrast, oxidized cholesterol solution can be stored at room temperature for many months without any apparent deleterious effect insofar as forming stable BLM is concerned.

e. In the process of forming BLM, gas bubbles may be introduced into the system, where they create optical and electrical interference, distort the membrane from planarity, and frequently cause rupture of the membrane. These bubbles arise from dissolved gases in the aqueous medium when its temperature is increased, or from air carried into the aqueous medium on brush, spatula, or syringe tip. Bubbles arising from the medium can be removed with a microsyringe or small spatula prior to membrane formation, and bubbles introduced during membrane formation may be avoided with sufficient care.

f. Temperature greatly affects initial membrane formation, membrane thinning, and BLM stability with any given BLM-forming solution. If a particular membrane-forming solution produces gray membranes which break in a fraction of a second, a few degrees centigrade increase in temperature may permit formation of stable membranes. If the initial membrane forms easily, but thins extremely slowly, a similar small increase in temperature may markedly accelerate the rate of thinning. Conversely, if a given BLM-forming solution rapidly forms BLM which rupture within a few seconds, a few degrees centigrade decrease in temperature may markedly increase the life expectancy of the BLM.

g. A thin lipid membrane that will not thin readily to the black state is often encountered. There are a number of ways to speed up the thinning

process. One can change the lipid solvent to one of a lower viscosity (e. g.,
dodecane to octane). Another effective method is to speed up the border suction by brushing the area around the aperture. A third, equally effective,
method is to poke the thin membrane with a fine object such as a wire or a
micropipet, thereby initiating the "zipper" process [25].

E. Characteristics of BLM Formation

1. Description of Characteristic Events

Direct optical methods are usually used to observe the formation of the
initial thick membrane, to follow subsequent thinning, to determine planarity
and continuity, and to measure the area of the BLM at equilibrium. When
the membrane is illuminated with white light and the light reflected from the
membrane is viewed through a low power microscope (magnification 10-40X),
three distinguishable stages may be observed during the thinning process.
These stages are schematically represented in Fig. 6.

The thick lipid membrane ($0.1-1.0 \mu$) which is initially formed appears
gray or colorless and may be considered to consist of a bulk hydrophobic
phase of membrane-forming solution with an adsorbed monolayer of lipid
molecules at each oil-water interface, i. e., at the biface. As the membrane
thins, interference color patterns appear and pass through several series of
higher-order bands of brilliant color to the first-order interference and
finally to a "silvery" golden appearance. At this point, the attractive forces
between the adsorbed lipid monolayers are still extremely small, and the
driving force for membrane thinning is the pressure differential due to presence of the Plateau-Gibbs border, i. e., the margin of bulk phase lipid in
which the membrane terminates. At the P-G border, the concave interfaces
create a pressure which is less than that in the interior of the thick planar
membrane. The P-G border therefore exerts a strong negative pressure
upon the interior of the thick portion of the membrane, causing bulk flow of
the interior phase and rapid reduction in membrane thickness.

The final stage is a completely transparent or "black" appearance which
develops with a sharply demarcated border from the "silvery" membrane as
the membrane thins to a bilayer (Fig. 6C). This stage is initiated with the
appearance of transparent or "black" spots (bimolecular thickness or small
multiples of bimolecular thickness) which result in a greatly increased rate
of thinning. A "zipper-like" mechanism has been suggested for the final
thinning process and is believed to be responsible for the rapid growth of
transparent or "black" area [47].

From classical optics it is well known that when light rays originate
from the same monochromatic point source, the light intensity is always
equal to the sum of intensities of the rays. In order to observe interference
phenomena, the superimposed waves must originate from the same point, as
shown in Fig. 7 (see Sec. III. A). If a thinning membrane is illuminated with
white light (an achromatic mixture of wavelengths), constructive and

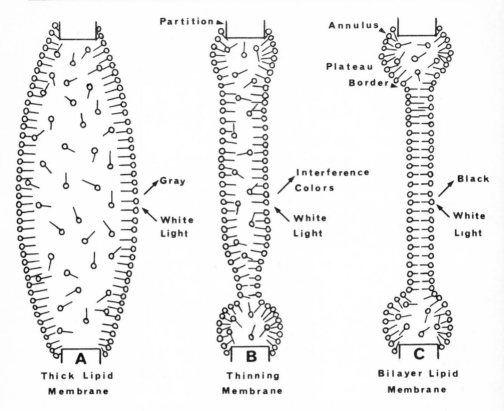

Fig. 6.   Diagrammatic representation illustrating the formation of a single and stable BLM.

destructive interference both operate to produce a sequence of chromatic bands of reflected light, termed interference colors or interference fringes. If a membrane illuminated with white light has thinned such that the optical path difference is equal to or slightly less than one-quarter of the average wavelength, constructive interference of the reflected light is maximal and the membrane appears golden yellow to "silvery" in color.

A membrane which is very thin relative to a quarter wavelength of the incident light exhibits essentially no reflection because of the complete destructive interference between reflective light wave fronts which are 180° out of phase. Such a nonreflecting membrane appears transparent or "black" rather than colored when viewed against a dark background. A faint gray shimmer of specular reflection of light from the "black" membrane may be seen under optimal conditions. It should be emphasized that such a membrane is not necessarily bimolecular or bilayer in thickness. Such "black" lipid membranes have been observed to undergo two, three, or more changes

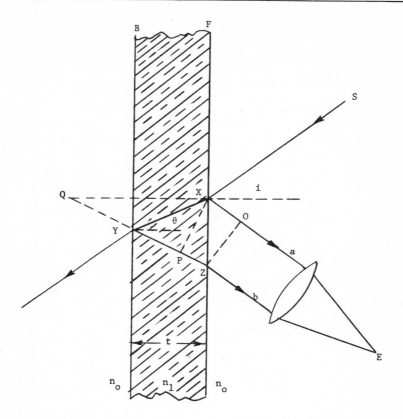

Fig. 7. Optics of a thin film. Shaded area represents a transparent ultrathin film of uniform thickness t and refractive index n_1 (or n_b) immersed in an aqueous solution of refractive index n_o (or n_w). S is a source of light. i and θ are the angles of incidence and refraction, respectively. a and b are interfering rays, superimposed at E, due to reflections at the two parallel interfaces F and B. OZ is the wave front of the reflected light.

in the degree of "blackness," with appropriate control of temperature and other conditions. The final stage of metastable equilibrium is, however, the stage of least reflection or greatest "blackness," often referred to as secondary black, and consistent with a bimolecular thickness as discussed in Secs. III and IV.

2. Drainage Types

Three different types of drainage have been observed as the initial thick membrane thins by movement of lipid into the surrounding bulk phase (the P-G border). The types of drainage observed during the formation of BLM in aqueous media correspond to the types of drainage described for aqueous soap films in air [25,48]. The major types are simple mobile, irregular mobile, and rigid drainage. In a BLM system under constant conditions, with the same aqueous medium, using the same membrane-forming method, etc., a given lipid solution generally exhibits predominantly a single type of drainage. However, the same solution can show more than one type of drainage and may show intermediate gradations between types, depending on the influence of physical factors such as changes in temperature or concentration. Such observations suggest that composition of the BLM-forming solution, the character of the particular lipids used, and external physical factors may all influence the type of drainage. They further suggest that during membrane thinning a process similar to a phase transition is occurring in the lipid phase.

Since the lipids used in forming BLM are generally amphiphilic, they tend to form a liquid crystalline phase when mixed in bulk with aqueous media [49]. A liquid crystalline phase exhibits simultaneously some fluid properties of a liquid and some properties of molecular order which approach the degree found in crystals. When the molar quantity of water is large relative to the quantity of amphiphilic lipid, a liquid crystalline phase consists of aggregations of amphiphilic lipid molecules which exist at a microscopic level as one of several forms of a phase separate from the bulk aqueous phase. Under these conditions, the hydrocarbon chains of the lipid molecules in each aggregate form a liquid phase separated from the bulk aqueous phase by an interface containing the hydrated polar portions of the lipid molecules. A liquid crystalline phase may exist as any one of a variety of transitional forms. Such forms range from simple micelles (aggregates of lipid molecules in spherical, cylindrical, and other configurations) to extended lamellar sheets. A transition of one of these forms to another may be caused by small changes in the composition, concentration, or temperature of a solution, or by external physical factors [50]. In the initial thick lipid membrane in aqueous media, the nature of the bulk hydrophobic phase is presently unknown, although there is some evidence that it is at least partially in a lamellar liquid crystalline phase.

The existence of three different transitional states, more or less analogous to three-dimensional solids, liquids, and gases, has been demonstrated in studies of lipid monolayers at the air-water interface. These have been called the condensed state, the expanded state, and the gaseous state. Such a monolayer behaves in two dimensions as a rigid, noncompressible solid in the condensed state, as a coherent but compressible fluid in the expanded state, and as an unordered gas in the gaseous state [22]. Changes in temperature, alterations in concentration of amphiphilic material in the spreading (monolayer-forming) solution, or changes in the length or degree of unsaturation of the hydrocarbon chains of the lipid can cause transition from one monolayer state to another [50].

The irregular mobile, simple mobile, and rigid types of drainage observed in thinning lipid membranes appear to be analogous, respectively, to the gaseous, expanded, and condensed states described for monolayers and to the transitional forms of the liquid crystalline phase.

In the simple mobile type, drainage is orderly, with smooth horizontal bands of color in the center of the membrane and somewhat more turbulence at the borders. As with all types, simple mobile drainage is recognizable only after the initial membrane has thinned sufficiently to exhibit interference colors. If the membrane-forming solution is sufficiently less dense than the aqueous medium, the lipid will predominantly move upward by bulk flow. The initial BLM formation will occur in the lower central portion of the membrane or at the lower margin. First-order colors will be exhibited from the black border upward in the sequence silver-blue, yellow, and rust-orange; second-order colors will follow in the sequence deep blue, green, yellow, orange, and red (Fig. 8). After such a membrane has thinned to greater than ca. 10% black, the boundary between silver and black areas remains essentially horizontal except at the margin. The rate of simple mobile drainage may vary from a very rapid rate requiring less than 30 sec to form a completely black membrane to a very slow rate such that the membrane has not become completely black at the end of 30 min.

Irregular mobile drainage is rapid, with complete black membranes forming in a few minutes. Initially, the horizontal color bands and movement of simple mobile drainage may be present. However, with the appearance of black areas, the membrane develops rapidly moving islands of thick and highly colored lipid which carry streamers of black with them in their ascent. Simultaneously, smaller areas of black may appear, rapidly descend, and coalesce with the lower portion of the membrane. The appearance of such a membrane exhibiting irregular mobile drainage has been likened to the appearance of peacock feathers. In the extreme case, turbulent swirling occurs throughout the plane of the membrane, with rapid appearance of multiple small black areas. Despite such turbulence, such membranes are generally stable and form BLM more rapidly and more regularly than any other type. Transition from simple mobile drainage to irregular mobile drainage is frequently observed at increased temperatures.

Very slow drainage with little or no turbulence, corresponding to the rigid type, has been observed less frequently. Interference fringes appear closely spaced, irregularly curved, and jagged-edged. Membrane-forming solutions and conditions which exhibit such drainage may also form membranes which do not appear to thin at all, but tend to persist with a gray appearance or to rupture with a "tear" rather than suddenly disappearing.

It should be emphasized that drainage patterns can be observed which are intermediate between the three major types, e. g., formation of a few "peacock feathers" in an otherwise simple mobile membrane, or a partially rigid and partially simple mobile membrane at a lower temperature.

A typical sequence for the appearance of a thinning membrane at various times after formation to the complete black state is shown in Figs. 8-11. This particular membrane exhibits predominantly a simple mobile type of drainage, with some elements of irregular mobile drainage.

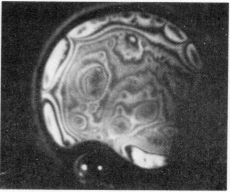

Fig. 8. The first of a sequence of photographs showing the formation of a BLM [25].

Fig. 9. Taken shortly after formation. Figures 8 and 9 are reproduced from frames of a color movie first shown at Symposium on the Plasma Membrane, organized by the American Heart Association and N. Y. Heart Association, New York City, December 8-9, 1961.

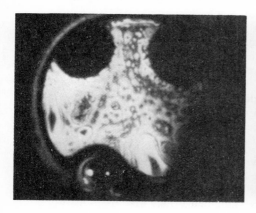
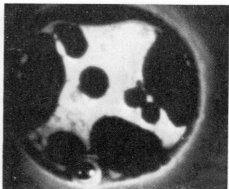

Fig. 10. Taken 10 min after formation. More than 25% of the membrane has become black.

Fig. 11. The growth of the membrane at a later stage.

Membrane-forming solution has drained upward and collected at the upper margin of the aperture, and multiple horizontal bands of interference colors have developed, 1.7 min after initial formation of the membrane (Fig. 8). A single circular black area is shown moving downward, with a trail of interference colors above it. The lowermost silver-yellow band is the thinnest nonblack portion of the membrane. The black portion of the membrane has increased in area and developed a horizontal border by 3.9 min after formation (Fig. 9). The large area of black in the lower portion of the membrane is intruded by two small peninsulas of thicker, colored lipid, and the remaining thick portions of the membrane have taken on more of the "peacock feather" appearance of irregular mobile drainage. By 6.4 min after formation, more than half the membrane area has become black, and the peninsulas of thicker lipid have disappeared from the bottom portion of the membrane (Fig. 10). The remaining colored portion of the membrane shows a further increase in turbulence with a central, upward-moving island of thicker lipid followed by a trail of black. Figure 11 shows the membrane 12.3 min after formation and 1.8 min after it has become completely black. The Plateau-Gibbs border where the BLM joins the bulk phase is distinct and only slightly irregular. In the lower half of the figure, a faint shimmer of reflected light from the otherwise transparent bilayer may be seen.

3. Rates and Mechanisms of Thinning

The rate of thinning of the membrane is defined most precisely as the net rate of volume change in the membrane with time. It is difficult to determine this rate of volume change directly, because it is a complex function of drainage and the dissolution of materials from the BLM-forming solution into the aqueous phase. Measurement of the increase in area of BLM with time is a more practical approach to the rate of thinning. Various methods have been used to measure the change in black membrane area and are discussed in greater detail in Sec. III. Thinning begins simultaneously with the formation of the initial membrane and proceeds continuously to the final black state. For any given membrane, the rate of thinning is constant, following the relationship $Tt^2 \cong k$, where T is the time, t is the membrane thickness, and k is the thinning constant [47]. Membranes which exhibit simple mobile drainage also exhibit a constant, although more rapid, rate of increase in black area. For BLM-forming solutions which yield such membranes under a constant and reproducible set of conditions, the rate of thinning seems to be characteristic. However, with irregular mobile drainage, the appearance of black spots results in a greatly increased rate of thinning.

The static and dynamic mechanisms of thinning described for soap films have also been observed in BLM formation. A static mechanism of thinning is a mechanism operative when the surface elements of the membrane remain in the same position with respect to the supporting aperture. Some evidence exists for the static mechanisms of stretching, viscous flow, and dissolution. The evidence for stretching is based on changes in black area of a planar membrane bowed by hydrostatic pressure and returned to a planar configuration. The area of black increases on bowing, and decreases to a size larger than that estimated from its normal rate of increase when the membrane is returned to planarity. Viscous flow occurs generally as bulk upward movement of liquid within a very thick membrane, which may be observed visually, with increased thickness of the P-G border at the top of the aperture. Dissolution during thinning is suggested by the extremely rapid thinning and frequent rupture of membranes formed with an aqueous-soluble solvent solution such as chloroform:methanol. In addition of hydrocarbon or other liquid additives will somewhat retard thinning and stabilize the membranes. The partitioning of solvents such as chloroform or methanol out of the membrane and into the aqueous phase is analogous to the evaporation of water from aqueous soap films in air.

Dynamic mechanisms of thinning refer to those mechanisms which involve motion of the membrane surface components with respect to the supporting aperture. The dynamic mechanisms of gravity convection and marginal regeneration are suggested by changes in interference color patterns which indicate the lipid elements in the membrane flow with respect to the stationary P-G border and aperture. In gravity convection, the thin elements of the membrane adjacent to the P-G border move in rapid downward currents, while the thicker central portion of the membrane moves slowly upward. In marginal regeneration, thicker membrane elements move into the P-G border as thinner membrane is pulled out of it in regularly spaced, immediately adjacent sections, giving a scalloped appearance.

III. MEASUREMENT OF MEMBRANE THICKNESS AND AREA

A. Optical Methods

The qualitative optical properties of thinning membranes during BLM formation were presented in Sec. II. E, with an introduction to the classical optics responsible for interference color formation. Optical interference methods are useful in semiquantitative measurement of membrane thickness and in the measurement of BLM area. As mentioned previously, the sequence of interference colors during membrane thinning passes through decreasing orders of interference color formation ending with the brilliant lower-order colors, which pass through a sharply demarcated transition zone to one, two, three, or more degrees of transparent or "black" membrane. Since the shortest wavelength of visible light is in the region $\lambda \simeq 4000$ Å, the final maximum of reflection intensity $(\lambda/4)$ occurs when the membrane thickness t is approximately 1000 Å. This indicates the upper limit for black membrane thickness, although the occurrence of three stages of black membrane would reduce this upper limit to about 300 Å. Therefore, although the considerations of interference optics do not permit quantitative determination of BLM thickness, they do provide a semiquantitative upper limit which is within the order of magnitude for BLM thickness determined by other methods.

As demonstrated in Fig. 11, the Plateau-Gibbs border reflects light strongly. Membrane area can thus be determined by direct visual or photographic measurement of the area enclosed by the P-G border, using a calibrated reticle at known magnification. Measurements of total membrane area, or of the area of black or of a particular color, can also be made directly from color photomicrographs. Measurements of black membrane area taken at known time intervals either visually or from photomicrographs may be used to calculate the rate of thinning or bilayer formation. It should be emphasized that measurement of the aperture size is not an adequate method of determining BLM area, because the amount and distribution of lipid in the P-G border may vary widely.

Quantitative determinations of membrane thickness may be achieved by measurements of BLM reflectance using theoretical considerations, assumptions, and experimental techniques originally derived from studies in thin, transparent solid or soap films. The optics of BLM belong to the general area of thin films (defined as a layer with parallel faces whose thickness t is of the order of the wavelength λ of light used) and specifically to the optics of ultrathin films, i.e., films with t less than 0.01 λ [47]. A BLM's reflectance is commonly measured at a small angle of incidence using incident light with a wavelength from the visible region of the spectrum. The BLM thickness is determined from the reflectance measurements using either a homogeneous single-layer theoretical model or an inhomogeneous triple-layer model. In the single-layer model it is necessary to assume that the membrane is homogeneous, isotropic, and transparent, and to either determine the refractive index of the membrane by measurements of the Brewster angle, or assume a value for the refractive index. In the triple-layer model, the polar portions of the lipid molecule situated at the biface are regarded as having a thickness t_p and a refractive index n_p, and the liquid hydrocarbon core is regarded as having a thickness t_h and a refractive index n_h. The overall BLM thickness t_m is thus equal to $(2\,t_p + t_h)$. Some aspects of theoretical background on thickness determinations by the reflectance method are given below.

1. Theoretical Considerations

The principle of this method is based upon the assumption that the optical properties of BLM in aqueous solution are fundamentally the same as those of ultrathin transparent solid or liquid films. Thus, when a beam of light is incident on a BLM of thickness t and refractive index n_b immersed in an aqueous solution of refractive index n_w, the character of light reflected from the front water-BLM interface differs from that of the black BLM-water interface (Fig. 7). The following specific assumptions are made:

a. The BLM is assumed to be transparent, homogeneous, and isotropic.

b. The reflected light is regarded as composed simply of rays a and b; the multiple reflections within the BLM are neglected.

c. The refractive index of the BLM is assumed either to be that obtained by observing reflectance at the Brewster angle, or that of the bulk material from which the BLM is generated.

The incident beam of light from source S is split into two parts upon reaching the water-BLM interface, F. The larger part of the two is the refracted ray which is again divided into two upon reaching the back interface, B. It is the reflected rays a and b which, after having traveled different paths, are superimposed to give the observed interference phenomena (at E). In short, the reflected intensity of a thin membrane in an aqueous solution depends on the refractive indices of the membrane and the aqueous phase, and on the path difference of the light rays in the membrane (i. e., on the thickness of the membrane, t). However, the refractive index increases at the front interface from n_w to n_b, and decreases at the back interface from n_b to n_w. The sign of the refractive index change is the same but opposite in sign for the interfaces, resulting in a 180° phase difference between the wave fronts of the first and second reflected ray. Additional phase difference between the two rays is introduced by the increased optical path traveled by the second ray; this increased path length is a function of the angle of incidence, the refractive indices involved, and the thickness of the membrane.

When the membrane thickness is equal to a quarter wavelength ($\lambda/4$) of normally incident monochromatic light of wavelength λ, the optical path difference of one-half wavelength ($\lambda/2$) produces an additional 180° phase shift in the second reflected ray. Constructive interference occurs under these conditions, causing a maximum intensity of reflection. Therefore, whenever the membrane thickness is such that the optical path difference is $\lambda/4$, $3\lambda/4$, $5\lambda/4$, . . . , there occurs a maximum of reflection intensity and a minimum of transmission intensity.

In the case of a BLM immersed in an aqueous solution, the resultant phase difference δ is given by

$$\delta = 2\pi n_b t_b \cos \theta / \lambda \tag{1}$$

where n_b is the refractive index of the BLM, t_b is the BLM thickness, θ is the refracted angle in the membrane, and λ is the wavelength of the light used.

It is only under the conditions stated above that the method of estimating thickness from the intensity of the reflected light can be simply applied. It can be shown that for the case of normal incidence when Eq. (1) is substituted into the Fresnel formula for the reflected amplitude E, in terms of r_1 and r_2, is given by

$$E = \frac{r_1 + r_2 \exp(-2i\delta)}{1 + r_1 r_2 \exp(-2i\delta)} \tag{2}$$

where $r_1 = (n_w - n_b)/(n_w + n_b)$, $r_2 = (n_b - n_w)/(n_b + n_w)$, $i = \sqrt{-1}$, and n_w is the refractive index of the aqueous phase, then the reflectance, R, defined as the ratio of the reflected energy I to the incident energy I_0 (i.e., E is multiplied by its conjugate E^*), is given by

$$R = \frac{I}{I_0} = \frac{r_1^2 + 2r_1 r_2 \cos 2\delta + r_2^2}{1 + 2r_1 r_2 \cos 2\delta + r_1^2 r_2^2} \tag{3}$$

Since the BLM in aqueous solution usually involves two identical interfaces, and therefore the reflection coefficients are equal except opposite in sign, Eq. (3) may be written in the following manner:

$$\frac{I}{I_0} = \frac{4r^2 \sin^2 \theta}{1 - 2r^2 + 4r^2 \sin^2 \theta + r^4} \tag{4}$$

For BLM in aqueous solution the Fresnel reflection coefficient, r, is quite small. Therefore, Eq. (4) reduces to

$$I/I_0 = 4r^2 \sin^2 \delta = 4r^2 \sin^2 [(2\pi n_b t_b \cos\theta)/\lambda] \tag{5}$$

The use of Eq. (5) instead of Eq. (4) in the calculation is well justified since the two equations differ by a factor $(1 - 2r^2 \cos 2\delta + r^4)$, which is seen to contain only the second and fourth powers of r. The error introduced by the approximation is negligible.

In Eq. (5), we see that the reflectance of the lipid film or membrane under consideration depends on the refractive indices of the film and the aqueous phase, and on the path difference of the light rays in the film (i.e., on the membrane thickness, t). The material thickness of the membrane may then be evaluated, provided we can measure the ratio of the light intensities and the refractive index of the membrane.

Since the fraction of the incident light reflected from the black membrane under study is estimated to be about 0.01%, the measurement of absolute reflectance presents a technical problem. Using the apparatus described in the experimental section, the reflected energy I from the black film can be measured without difficulty. However, this entails the use of a high intensity source, since I_0 is about four orders of magnitude higher than I. To avoid the direct measurement of I_0, the usual practice used in soap film studies is followed. For membranes still exhibiting interference fringes, the thickness of the membrane is given by

$$t = k\lambda/4n_b \cos \theta \tag{6}$$

where k is the order of interference. When $k = 1$ the membrane gives the maximum reflection in the silvery state before the transition to the black state. Therefore, Eq. (5) may be written for the two cases

$$I_s/I_0 = 4r_s^2 \sin^2 \delta_s \tag{7}$$

for the membrane with the maximum reflection, and

$$I_b/I_o = 4 r_b^2 \sin^2 \delta_b \tag{8}$$

for the BLM, where the subscripts s and b refer to the membrane in the silvery-golden and black state, respectively. The reflectance ratios, Eqs. (1), (7), and (8), may be combined to give

$$\frac{I_b}{I_s} = \left(\frac{r_b}{r_s} \right)^2 \sin^2 \frac{2 \pi n_b t_b \cos \theta}{\lambda} \tag{9}$$

where r_b and r_s are the Fresnel coefficients for the black and silvery films, respectively. The ratio I_b/I_s is measured at wavelength λ.

As a first approximation of BLM thickness, Eq. (5) is quite adequate for most purposes. However, it has been pointed out by Tien [51] that a BLM is not isotropic in its properties. This is owing to the fact that substances which can form BLM so far established are all amphipathic compounds. Hence, the BLM may be regarded as a symmetric triple-layered structure, as illustrated in Fig. 12. In this case the central hydrocarbon layer has a different refractive index from the layers situated at the biface where the hydrophilic groups are located. For the triple-layered model light approaches from the various interfaces from both directions, except the last interface (Fig. 12).

The essential steps in the derivation of the equation as applied to BLM in aqueous media have been given [51]. The final result is

$$R = \frac{r_1^2 + r''^2 + 2r_1 r'' \cos (x - \delta_h)}{1 + r_1^2 r''^2 + 2r_1 r'' \cos (x_1 - \delta_h)} \tag{10}$$

where r's are the Fresnel reflection coefficients, which are given by the formulas:

$$r_1 = -r_4 = \frac{n_p - n_w}{n_p + n_w} \tag{11}$$

$$r_2 = -r_3 = \frac{n_h - n_p}{n_h + n_p} \tag{12}$$

The phase differences x, the resultant phase change, and reflectances r' and r'' are given by

$$x_1 = x_3 = \frac{4 \pi n_p t_p \cos \delta}{\lambda} \tag{13}$$

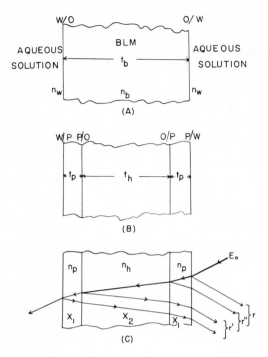

Fig. 12. Models of BLM used in thickness calculation. (A) Single-layered model, (B) triple-layered model, (C) reflection of light from the triple-layered membrane model, E_o is incident light wave, X's refer to phase difference due to different layer. r, r', and r'' are amplitudes of the reflected light at various interfaces (see text for details).

$$x_2 = \frac{4\pi n_h t_h \cos \theta}{\lambda} \tag{14}$$

$$r'^2 = \frac{r_3^2 + r_4^2 + 2r_3 r_4 \cos x_1}{1 + r_3^2 r_4^2 + 2r_3 r_4 \cos x_1} \tag{15}$$

$$r''^2 = \frac{r_2^2 + r'^2 + 2r_2 r' \cos (x_2 - \delta_p)}{1 + r_2^2 r'^2 + 2r_2 r' \cos (x_2 - \delta_p)} \tag{16}$$

In the equations given above, n_w is the refractive index of the aqueous medium, t is thickness, λ is the wavelength of light, and δ and θ are the refracted angles in the polar and hydrocarbon regions, respectively. Although these equations are cumbersome to apply without the aid of a computer, they essentially depend only on a reflectance measurement at some known angle and a knowledge of the refractive index of each region. At present, the parameters which cannot be directly measured are n_h, t_p, t_h, and n_p.

2. The Brewster Angle of BLM

The refractive index of the film is determined from measurements of the Brewster angle. This method is based on the fact that for light polarized with the electric vector in the plane of incidence, the reflectance of a film of refractive index n_b immersed in a medium of index n_w at the angle is defined by $\tan \varphi_p = n_b / n_w$, where φ_p is the so-called polarizing angle. In terms of the Fresnel coefficient for the reflectance of light with this plane of polarization, we have

$$r = \frac{[\tan (\varphi - \theta)]}{[\tan (\varphi - \theta)]} \tag{17}$$

where φ is the angle of incidence and θ is the refracted angle in the BLM. When $r = 0$,

$$\tan (\varphi + \theta) = \infty \tag{18}$$

$$\varphi_p + \theta = 90^\circ \tag{19}$$

It follows from Snell's law that

$$n_w \sin \varphi_p = n_b \sin \theta \tag{20}$$

and therefore

$$\tan \varphi_p = n_b / n_w \tag{21}$$

This simple method involves practically no calculation and is only dependent on a knowledge of the refractive index of the medium. This can be easily determined. Further, no absolute intensity measurements are required if one is only interested in the refractive index.

Application of Brewster's law to a triple-layered BLM is complicated. As is shown for the single-layered model above, there is also a unique angle of incidence for the triple-layered model for which the reflectance of polarized light in the plane of incidence is zero. However, the value of this angle is a function of the thicknesses and refractive indices of the various layers. Since there is some uncertainty regarding the application of Brewster's law even to a single-layered BLM, it is of dubious value to consider this problem further. Interested readers are referred to a number of standard texts on thin film optics [52, 53].

3. Apparatus and Procedures

The optical methods used are essentially the same as those devised by
workers in the field of soap film studies [48, 54, 55]. The cell used for membrane formation and reflectance measurements is shown in Fig. 13.

The BLM are formed on a small loop made of Teflon (4 mm in diameter,
manufactured by Chicago Gasket Co.). This loop (or O-ring) is positioned at
the end of a capillary tube which has been enlarged and is held in place by
means of two platinum wires which are embedded into the glass. The capillary tube is fixed in a ground-glass joint (size 12/30; other types of joint,
such as Solv-seal or ball, are also suitable). The other end of the capillary
tube is connected to polyethylene or Teflon tubing (1 mm or less i.d.) which
in turn is connected to a syringe via a three-way valve (Hamilton Co.). This
syringe is moved by an infusion-withdrawal pump (Harvard Apparatus Co. ,
Model No. 600-000), the three-way valve facilitating the filling of the syringe
without its removal from the pump.

The cell chamber into which the loop is immersed may be jacketed for
temperature control if desired.

The lipid solution for membrane formation is introduced by means of
the infusion-withdrawal pump. Openings are provided for the inner cell so
that the contaminated aqueous solution can be renewed frequently (see Fig. 13).
The whole cell assembly, with the exception of a porthole for light path, is
painted black to minimize stray reflections.

Any intense and stable white light source may be used in combination
with an interference filter. A microscope illuminator (American Optical Co.)
is used with a Bausch and Lomb selected line filter (5461A) in the manner
illustrated in Fig. 13. The light from the source is passed through the pin-
hole, the interference filter, and two positive lenses which focus it to a small
spot on the center of the BLM. The front of the cell is tilted slightly so that
any light reflected from the glass outer chamber is not seen by the viewing
microscope. A photomultiplier tube (RCA 1P21 or EM-9558Q) attached to a
third eyepiece on the microscope is used to measure the intensity of the reflected light from the membrane. (In this instance the third eyepiece is
fitted to a low power stereomicroscope, A-O Spencer Cycloptic 59 F#D2, by
attaching a photographic tube adapter, all made by American Optical Co.)
With this third eyepiece, the membrane can be observed visually during
thinning. The output from the photomultiplier tube is measured by a micro-
photometer (American Instrument Co.) and the signal may be recorded with
the aid of a strip chart recorder. All of the optical components should be
mounted rigidly on optical benches. If a dark room is not available, the cell
assembly and optical components should be enclosed in a large box with the
inside painted black. The box, with a sliding panel (or heavy curtain) on one
side, provides an easy access to the setup and at the same time permits

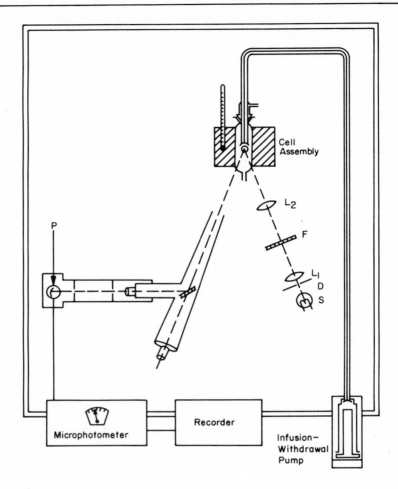

Fig. 13. Apparatus for measurement of reflectance from a BLM:
S-light source, D-iris diaphragm, L_1-condensing lens, F-interference filter,
L_2-focusing lens, P-photomultiplier tube (RCA 1P21 or EMI-9558 Q).

experiments to be carried out in the dark. It should be mentioned that the
angle between the optical benches should be as small as possible (less than
$20°$) so that an essentially normal incidence of light is achieved.

In the above arrangement the optical benches may be dispensed with,
however, if a goniometer table is available. Also, a He-Ne laser may be
used as a monochromatic light source.

Essentially the same setup can be used in the determination of Brewster's
angle. Plane-polarized light is produced by inserting a piece of Polaroid film
(or its equivalent) in the incident light path. Owing to the drastic reduction

in light intensity, the use of an interference filter is not usually possible
with white light.

The following variables should be kept in mind in designing a workable
optical setup:

a. A light source of sufficient intensity must be used, since the BLM
reflects a very small fraction of the incident light.

b. A very small spot of light must be incident on the BLM proper so that
no light is reflected from the P-G border (see Sec. II) or from the loop itself.

c. There must be no interference with the light reflected from the cell
chamber.

d. Means should be provided for orienting the BLM relative to the light
beam.

e. Means must be provided for visual observation of the membrane so
that one can be certain there are no air bubbles or solid particles, etc.,
which would give a spurious reflectance measurement.

The operating procedure for reflectance measurements is as follows:

a. To form a BLM, the syringe is filled with a BLM-forming solution
which is pumped at a fairly fast rate onto the loop via the capillary tube
(see Fig. 13).

b. After a globule is formed which envelopes the opening of the loop,
the infusion-withdrawal pump is reversed to withdraw the excess BLM-
forming solution from the loop.

c. The pump is stopped when a thin film is seen on the loop.

d. Reflectance of the membrane may be monitored during the above
steps. A typical trace from the initial thick membrane to the final black
state is shown in Fig. 14 (taken from Ref. [56]).

It should be mentioned that there is an optimal size for the Teflon loop.
When a BLM-forming solution possesses a low interfacial tension (see Sec.
VI), it is extremely difficult to cover the loop. To circumvent this difficulty
the injection or interface application method may be used (Sec. II.A).

For Brewster angle measurements, the pivoted optical benches are rotated and the angle between the incident and reflected light is measured. A
goniometer is most suitable for this measurement. The position at which
the intensity of the reflected light is zero may be detected either visually or
with the aid of a photomultiplier. It should be noted that, owing to light refraction by the aqueous phase, a small correction of the measured angle
between the incident and reflected light is necessary. The corrected angle
is given by

$$\varphi_p = 45 + [(\varphi' - 90)/2n_w] \tag{22}$$

where φ' is the measured angle between the light beams and n_w is the refractive index of the aqueous solution.

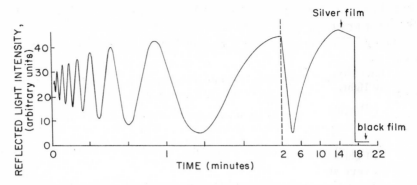

Fig. 14. Strip chart recorder tracing illustrating the formation of a thin lipid membrane under monochromatic light (5461 Å) and its thinning to bilayer thickness [56].

4. Interpretation of Optical Data

To date, only one systematic study on a BLM system formed from glycerol dioleate has been carried out using optical methods [56]. We shall use the results obtained [56] for our discussion and to illustrate the significance of the optical measurements in relation to the BLM structure.

The BLM-forming solution used in the systematic study was simply prepared by mixing 0.4 ml of glycerol dioleate (K & K Rare Chemicals Co.)with 100 ml of n-octane (Eastman Kodak Co.).

a. Thickness Measurements. The thickness of BLM is calculated from reflectance data using the classical Rayleigh equation given earlier [Eq. (5)]. In actual practice, Eq. (9) is generally used. The membrane in the silvery state (i.e., the state of maximum reflection before going black) produced from a 10% solution of glycerol dioleate in n-dodecane formed in 0.1 N NaCl at 30° C is used to measure quantity I_s [see Eq. (9)]. The intensity, I_b, of light reflected from BLM is easily measured with the apparatus described in Sec. III.A.3 (see Figs. 3, 13). To calculate the BLM thickness, t_m, from Eq. (9), the refractive indices of the membrane in the silvery and black state must also be known. These values are obtained from experimentally measured Brewster angles and values of n calculated with the aid of Eq. (21). The refractive index of the membrane in the silvery state is 1.43 ± 0.02 (n_s), whereas the value obtained for the black membrane is 1.48 ± 0.03 (n_b). These refractive indices are used in the thickness calculations. For a comparison, the results of refractive indices obtained by a number of workers are given in Table III.

b. The Effect of Electrolyte Concentration on BLM Thickness. A number of experiments have been carried out to study the effect of salt concentration and the nature of salts on BLM thickness. The results of these experiments are given in Table IV. It can be seen that for the system studied, the BLM

TABLE III

Refractive Index Values of Thin and Black Membranes in Aqueous Media

Lipid Solution	State of membrane	Temp, °C	Refractive index	Ref.
Egg lecithin in $CHCl_3$ + $CH_3 OH$ + tetradecane	Silvery	36	1.49 ± 0.03	[37]
Egg lecithin in $CHCl_3$ + $CH_3 OH$ + tetradecane	Black	36	1.66 ± 0.03	[37]
Glycerol distearate in n-octane	Silvery	32	1.42 ± 0.03	[25]
Glycerol distearate in n-octane	Black	32	1.56 ± 0.03	[25]
Egg lecithin in n-decane	Silvery	30	1.44 ± 0.03	[51]
Egg lecithin in n-decane	Black	30	1.60 ± 0.03	[51]
Egg lecithin in n-decane + tetradecane	Black	20	1.37 ± 0.02	[57]
Glycerol dioleate in dodecane	Silvery	30	1.43 ± 0.02	[56]
Glycerol dioleate in n-octane	Black	5	1.48 ± 0.03	[56]

TABLE IV

The Effect of Electrolyte Concentration and Various Electrolytes
on BLM Thickness [a]

Electrolyte	Concentration (molarity)	Refractive index of aqueous phase	Thickness, Å
Distilled H_2O	-----	1.334	53
NaCl	0.001	1.334	54
NaCl	0.01	1.334	53
NaCl	0.1	1.335	51
NaCl	1.0	1.346	51
NaCl	4.0	1.379	50
KCl	0.1	1.335	50
NaI	0.1	1.336	52
Na_2SO_4	0.1	1.336	53
$CaCl_2$	0.1	1.337	52

[a] All BLM formed from a 0.4% solution of glycerol dioleate in
n-octane at 5°C.

thickness remains practically constant over a wide concentration range and
also independent of the type of electrolytes used [56].

c. The Effect of pH of the Aqueous Phase on BLM Thickness.   The pH
of distilled water was varied from 0.9 to 11.5 by adding either HCl or NaOH.
At pH 12 the lipid solution would adhere to the loop but would not thin down to
even a thick color membrane.   The results of thickness measurements as a
function of pH are shown in Fig. 15.   It can be seen that for the system
examined there is little change in thickness from pH 0.9-9.5.   Above pH 9.5,
the thickness increases rapidly as pH of the aqueous solution is further increased.   Similar results were obtained when the BLM were formed in 0.1 M
NaCl.   However, in the latter case the pH range examined was 4.7-8.5,
since the lipid solution would not thin above pH 9.0 in 0.1 M NaCl [56].

d. The Effect of Temperature on BLM Thickness.   For glycerol dioleate
in octane, stable BLM can be formed in the temperature range 0 to about
15° C.   Measurement of the reflectance over this temperature range did not

show an observable change.   The available data are too limited for us to
draw any definite conclusions concerning the temperature effect.

    e.   Thickness Evaluation: Single- vs. Triple-Layered Model.   In the
above thickness calculation, the explicit assumptions are that the BLM is
transparent, homogeneous, and isotropic (Sec. III. A. 1).   As has been discussed above, the fact that amphiphilic compounds are necessary in producing stable BLM suggests a triple-layered structure (Fig. 12) which corresponds
more closely to the reality of the membrane.   In the case of the triple-layered
model, the refractive index of the inner hydrocarbon core is different from
that of the outer regions (hydrophilic layers).   At present there is no known
experimental method which will permit an assessment of the refractive index
of the inner layer of a BLM.   Therefore, in order to use Eq. (10) one is
compelled to use an assumed value.   An analysis using the triple-layered
model has been made [51], which from the calculated thickness values seems
to indicate that the hydrophilic groups at the biface should be fully extended
in the aqueous phase.   In general, the thickness of a BLM calculated according to the triple-layered model [Eq. (10)] is about 10% higher than that calculated using the single-layered model [Eq. (5)].

B.   Electrical Methods

    BLM thickness and area may be estimated by electrical methods based on
the measurement of transmembrane capacity, provided that certain assumptions are made about the dielectric character and thickness of the membrane.
The determination of membrane thickness from measurements of membrane

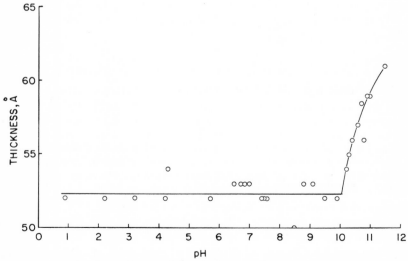

Fig. 15.   The dependence of the thickness of a BLM formed from a
0.4% solution of glycerol dioleate in n-octane as a function of the pH of the
aqueous phase [56].

capacitance is less precise than optical reflectance methods but is experimentally and mathematically more simple and more rapid, and thus provides a useful check on the extent to which a particular BLM is in fact of bilayer thickness. The monitoring of BLM area is achieved by measurement of transmembrane impedance with low frequency alternating current and depends on the essential constancy of normalized bilayer capacitance. The detailed description of apparatus and methods for determining membrane capacitance and membrane impedance are presented in Sec. IV, and only the theoretical and mathematical arguments are presented here.

It is generally assumed that a BLM system (aqueous solution-BLM-aqueous solution) can be represented by the equivalent circuit shown in Fig. 16. The capacitance of the BLM has been found to be independent of frequency in the range from 5×10^{-3} to about 10^7 cps, directly proportional to the BLM area, and apparently dependent only on the dielectric layer of the hydrophobic compounds in the membrane (see Sec. IV). The capacitance of the membrane according to the parallel-plate condenser formula is given by

$$C_m = \frac{\varepsilon A}{4\pi t_d} \simeq \frac{8.8 \varepsilon A}{t_d} \qquad (23)$$

where C_m is the BLM capacitance in μF, A is the BLM area (cm^2), ε is the dielectric constant, t_d is the thickness of the dielectric layer in Angstrom units, and 8.8 is the conversion factor. Therefore, Eq. (23) may be used to determine one of the parameters when the others are known or assumed. For instance, the thickness t_d of the BLM can be evaluated from a knowledge of the measured capacitance C_m and BLM area together with an assumed dielectric constant ε. The dielectric constant of unmodified BLM is usually assumed to be between 2 and 4, since the dielectric constants for most liquid hydrocarbons lie in this range. For a circular BLM, the area is given by

$$A = \frac{\pi D^2}{4} = \frac{C_m t_d}{8.8 \varepsilon} \qquad (24)$$

where D is the diameter of the membrane (see Sec. VI).

The working equation for the estimation of the BLM thickness is

$$t_m = 2t_p + t_h \simeq 2t_p + t_d \qquad (25)$$

where t_p is the thickness of the hydrophilic portion of the membrane, t_h is the thickness of the hydrocarbon interior of the membrane, and t_d is the calculated thickness of the dielectric portion of the membrane, which is assumed to equal the thickness of the hydrocarbon portion.

An extended molecule of phosphatidyl choline has a calculated (from models) overall length of 34.5 Å and a length of about 25 Å for the hydrocarbon chains of the fatty acids [4]. A BLM formed from phosphatidyl choline should thus have an overall thickness of about 70 Å and a thickness t_d of about 50 Å, assuming no interdigitation of the hydrocarbon chains.

It is assumed in this method that only the hydrocarbon region $t_h \simeq t_d$ is involved in BLM capacitance, and that the film behaves essentially as a

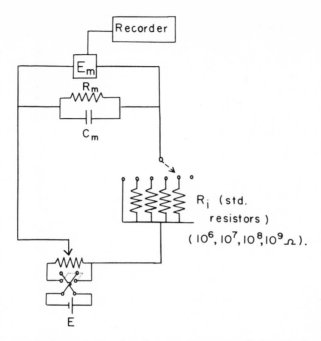

Fig. 16. Equivalent circuit for dc measurement of electrical properties of BLM. E_m-electrometer, R_i-input resistors, C_m-membrane capacitance, R_m-membrane resistance, E-polarizing potential source.

parallel-plate condenser. In addition, some value must be assumed for the thickness of the polar group region (t_p), e.g., $10\,\text{Å}$. It is further assumed that the permeativity, ε, of the effective dielectric portion of the membrane is similar or equal to the permeativity values measured by electrical or optical methods for bulk hydrocarbon.

An electrical method for monitoring BLM area has been developed [29] which provides a continuous and automatic record of membrane formation, thinning, increase in BLM area, BLM duration, and eventual rupture. ·The method is made possible by the essential constancy of the BLM capacity and the fact that membrane impedance, Z, becomes an increasingly pure function of the capacity as the alternating current frequency increases above $10/\tau$, where the time constant τ may be measured from the voltage vs. time plot (see below).

The technique consists of applying an appropriate sinusoidal, e.g., 1000 Hz, signal across the membrane through a current-limiting resistor, detecting the transmembrane potential difference with a high input-impedance electrometer preamplifier, amplifying the capacitance-coupled signal with an ac amplifier, rectifying it, and recording the rectified signal on a strip chart recorder. Such a recording is shown in Fig. 17, in which the ordinate is an

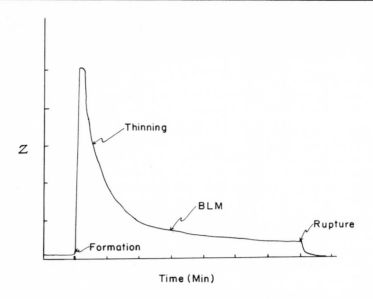

Fig. 17. Strip chart recorder tracing indicating the formation of a lipid membrane, its thinning to bilayer thickness, and its rupture. The ordinate is a nonlinear function of the membrane impedance Z.

exponential function of the membrane impedance, reflecting primarily the changes in capacitive reactance of the membrane at various stages. Although the formation and initial stages of thinning of a lipid membrane in aqueous media are not related in any simple mathematical manner to such a record, the initial appearance, growth, duration, and final rupture of BLM are directly related to the method. Because this method provides a continuous and automatic record of BLM life history, it is particularly useful when the investigator is performing other operations such as the addition of solutes or withdrawal of samples necessary in permeability studies.

C. Electron Microscopy

Although the development of primary and secondary black spots in the interference color patterns of thinning lipid membranes provided a semi-quantitative estimate of the BLM thickness, the first quantitative estimations of thicknesses were made from electron micrographs [24]. The OsO_4 technique used in these studies is described in a later paragraph. The published electron micrographs of stained BLM demonstrated a thickness of 60-90 Å, with an occasional suggestion of the trilamellar dense-light-dense appearance now well established for biological membranes [8, 58, 59]. The initial workers used OsO_4-fixed and -stained BLM preparations.

Further electron microscopic studies on BLM were carried out by Henn et al. [60], who visualized BLM both as thin sections and as shadowed

preparations. These workers fixed the BLM with lanthanum nitrate and potassium permanganate, air-dried the fixed BLM, and embedded them directly
in an epoxy resin before sectioning and study by electron microscopy.

1. Practical Details

The essential steps in obtaining electron micrographs of a BLM are
listed below. Here a BLM is treated as a thin, biological tissue section.

a. _Membrane Formation._ A BLM is formed in the usual manner either
on a hair loop or on a support made of Lucite (Plexiglas) (Sec. II. C). The
diameter of the BLM should be kept about 2 mm or less to enhance its stability
and subsequent handling. The use of Plexiglas instead of polyethylene or
Teflon is said to give better adhesion between the embedding resin and the
BLM support.

b. _Membrane Fixation._ Typical fixatives such as OsO_4 , $KMnO_4$, and
$La(NO_3)_3$ have been used with varying degrees of success. In the case of
OsO_4 fixing, a BLM is formed on a hair loop in a large vessel containing two
small cups. One cup contains buffered OsO_4 and the other 3% agar solution.
The BLM is formed on the hair loop and then transferred into the OsO_4 cup
for 10-20 min, then immersed in the agar cup. The entire process is carried out under 0.1 M NaCl at 40° C.

According to Henn et al. [60] , only $La(NO_3)_3$ followed by $KMnO_4$ treatment permits the formation of a BLM of sufficient stability for further
processing. The procedure of using $La(NO_3)_3$ - $KMnO_4$ fixing is as follows:
A BLM is formed on a hole (1.5 mm) in a thin Plexiglas square immersed in
9 ml of 0.1 M NaCl at 36° C. To the well-stirred solution, 1 ml of 10%
$La(NO_3)_3$ is added. This is followed by adding 2 ml of saturated $KMnO_4$
solution. The fixed BLM can be removed from the solution and air-dried.

c. _Embedding and Sectioning for Electron Microscopy_. In the procedure
of Mueller et al. the BLM is embedded in Vestopal resin after first hardening
the agar gel by lowering the temperature and dehydrating the gel with alcohol. The preparation is sectioned with a microtome in the standard way.

If Epon 812 resin is used, the dehydration step is found to be unnecessary
[60] . Here the air-dried preparations are embedded directly in the resin
and polymerized at 60°C overnight.

d. _Electron Microscopy_. Localization and positive identification of
BLM structures are obtained by searching grids at low magnification until a
segment of the BLM support and attached thick membrane (i.e. , the P-G
border) are found. The structure is then followed over its entire length at
high magnification. An RCA electron microscope was used by earlier workers.
Henn et al. [60] used a Siemens Elmiskop microscope operated at 80 kV.
The microscope was calibrated by means of a diffraction replica grating.
Micrographs were obtained with optical magnifications ranging from 4000 to
40,000X. The BLM thickness was estimated by comparing measured lengths
of the shadows of tomato bushy stunt virus (300 $\pm$ 10 Å in diameter).

2. Results and Interpretation

Figure 18 shows an electron micrograph of a BLM formed from a mixture of egg lecithin and cholesterol (molar ratio 10 : 1) at 118,000X magnification (H. A. Blough and G. Gordon, private communication). In the central portion of the picture, multilamellar structures are discernible. The micrographs published by earlier workers [24] indicate that the width of well-stained preparations is about 60-90 Å with an occasional suggestion of double lines. The micrographs published by Henn et al. [60] give a clear-cut image of a trilamellar structure, although their results show a wide variation in thickness. Peak-to-peak distances ranged from 37.5 to 116 Å as determined from microdensitometer tracings, with a mean BLM thickness of 73.4 + 21.5 Å in terms of peak-to-peak distance. This value is in reasonable agreement with the determinations of BLM thickness by other methods (Sec. III. A, B).

Thus far, electron microscopy of BLM has provided additional evidence that the structure of BLM is similar to myelinic figures exhibited by phospholipids extracted from the brain when placed in water [59]. It is interesting to note that when myelin figures are developed in the presence of a basic protein (e. g., globin) the overall width of the tripartite structure is increased by 25-50 Å, whereas the central region (unstainable hydrocarbon) remains the same (i. e., about 20 Å). Whether a corresponding increase in thickness would also be observed when a BLM is similarly treated remains to be demonstrated.

In concluding this section on measurements of BLM thickness and area by various methods, it is relevant to point out (a) that electron microscopic studies of BLM are subjected to the usual uncertainty of staining and embedding steps, (b) that the electrical capacity method, on the other hand, requires the use of an assumed dielectric constant in the calculation, and also that the presence of an electric field across the membrane may alter the structure of the BLM, and (c) that the drawback of the optical methods lies in the evaluation of the refractive index of the BLM. The advantages of the optical methods are that the BLM structure is least disturbed during measurements and that interactions between the BLM and protein (or other compounds) may be studied in situ.

IV. MEASUREMENT OF ELECTRICAL PROPERTIES

A. General Considerations

Electrical measurements are the methods most used in characterizing the properties of BLM and provide simple means to monitor changes and interactions between the BLM and its modifiers (any agent which alters the basic properties of the BLM is defined as a "modifier"). Both ac and dc methods have been used for the measurements of various electrical parameters. Because of the very low conductance of unmodified BLM, great care is needed in the insulation of membrane chamber, electrodes, switches, and

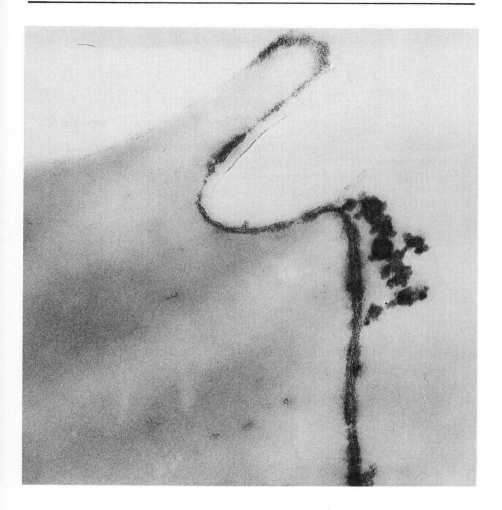

Fig. 18. Electron micrograph of a BLM formed from a mixture of lecithin and cholesterol (molar ratio 10 : 1) showing multilamellar structures. The BLM was imbedded in Epon resin following fixation with 12% $KMnO_4$ (courtesy of Dr. H. A. Blough). Magnification 118,000X.

connections to avoid current leakage. It is generally advisable to enclose the whole setup in a cage made of copper screen to avoid electrostatic interferences.

The general arrangement of the cell for measurement of electrical properties may be represented as follows:

| nonpolarizable electrode | aqueous solution | BLM on polyethylene or Teflon support | aqueous solution | nonpolarizable electrode | (26) |

The aqueous solutions serve essentially as ohmic contacts to the membrane.

B. Apparatus and Circuitry

1. Cell Assembly

To facilitate cleaning and ease of manipulation, a simple cell assembly is recommended. This consists of a glass outer chamber such as that shown in Fig. 20. The inner chamber can be made from a polyethylene pH cup (5-ml, Beckman) or a Teflon beaker (10 ml, commercially available). The small aperture (or hole) of about 0.5- to 2-mm diameter is produced in the side of the polyethylene cup as follows: The side of the pH cup is first thinned to about 0.2-mm thickness by pressing it with heated pliers acting over aluminum foil (to prevent sticking); then a heated needle held in a drill chuck is used to bore a polished hole in the thinned portion of the wall. In the case of a Teflon beaker, a portion of the wall is first machined down to about 0.025 cm and a small hole (1-2 mm in diameter) is punched through it. Whenever available, Teflon is preferred over polyethylene for the inner chamber, because the former not only provides a highly hydrophobic substrate for the BLM-forming solution but is also chemically inert and structurally less deformable.

2. Electrodes

Several types of "nonpolarizable" electrodes have been used, including saturated calomel, silver-silver chloride (Ag-AgCl), bright platinum (Pt), and platinum coated with platinum black. Commercially available saturated calomel or Ag-AgCl electrodes are most easily connected to the aqueous phases via bridges made of Teflon or polyethylene tubing containing 2% agar saturated with KCl. Calomel electrodes (such as Beckman No. 41239), when properly prepared and maintained, are effectively reversible and nonpolarizable, introducing an emf of less than 1 mV into the measuring circuit. Introduction of electrode emf and tendency to polarization increase in the order: Ag-AgCl, platinum coated with platinum black, and bright platinum.

3. Other Provisions

The temperature of the cell assembly may be maintained either by immersing the cell in a constant temperature bath or by circulating water through a glass coil (placed around the inner chamber) from a thermostatic bath (Model F, Haake). For effective stirring, a magnetic stirrer may be used (Fig. 19). If required, the solution in one or both chambers may be changed with the BLM in place by using infusion-withdrawal pumps. A pair of closely matched syringes should be used. To prevent the membrane from bulging due to small differences in the volumes exchanged, an additional syringe operated by hand may be added. The syringe pumps manufactured by Harvard Apparatus Co. (Dover, Mass.) have been found suitable.

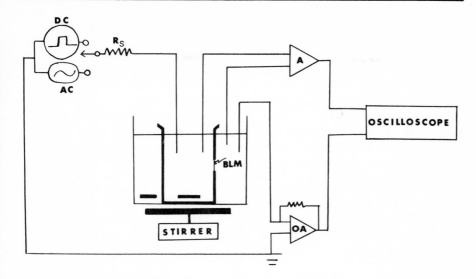

Fig. 19. Schematic diagram showing an experimental arrangement. The current from either an ac or a dc source is made to flow through a series resistor (R_S) across the BLM. The current electrode in the inner compartment is connected to the summing point of an operational amplifier (OA). The voltage changes (V) across the BLM are differentially recorded with the aid of a high input impedance amplifier (A). Both voltage V and current I are displayed on a dual-beam cathode ray oscilloscope (such as Tektronix 502). The solutions in both chambers are stirred magnetically.

4. Restrictions

The extremely high resistance of unmodified BLM imposes a number of restrictions on the quality and types of electrical circuit components and their connections. The foremost of these restrictions is that the insulation of all components above ground potential, such as switches, plugs, variable resistors, etc., must provide a leakage resistance not less than $1 \times 10^{10} \, \Omega$ to ground at any point. Thin, invisible films from body oils, machining oils, laboratory atmosphere, etc., may provide leakage paths with resistances many orders of magnitude lower than this. Great care in assembling and handling circuit components and in making connections is necessary, and contaminating, conductive films may generally be removed by careful swabbing with alcohol.

A second set of restrictions caused by high resistance in BLM systems is that of low stray capacitance and low degrees of capacitative coupling. Stray capacitance arising in electrode leads, electrodes, chamber apparatus, and occasionally within the electronic instruments utilized in the measurements may easily exceed the fraction of a microfarad capacitance of the BLM itself and lead to significant errors in measured values. Capacitative

164   H. TI TIEN AND R. E. HOWARD

coupling primarily occurs between wire leads and becomes increasingly troublesome when ac methods are used at higher frequencies. Use of low noise, shielded cable with thick insulation between the core conductor and the grounded shield, and careful immobilization of leads, will generally reduce capacitative coupling to acceptable levels even with ac methods.

A third set of restrictions concerns the electrical components and instruments used for measurement of high resistance BLM properties under usual constant current conditions. Under ideal conditions the output impedance of current signals should be greater than 10 times the BLM resistance on both sides of the membrane, and the input impedance of the voltage-measuring components should likewise be greater than 10 times the BLM resistance on both sides of the membrane. In usual practice one side of the BLM system is at a true or effective ground potential, and the current output impedance and voltage input impedance are maintained at high values only on the nonground side of the system.

Figure 19 shows an equivalent electrical circuit in which one side of the BLM is at ground potential. This circuit may be assembled from widely available electronic components and is suitable for general characterization of dc electrical properties of BLM. The critical components of such a system are the current-limiting resistance R_s and the electrometer voltmeter.

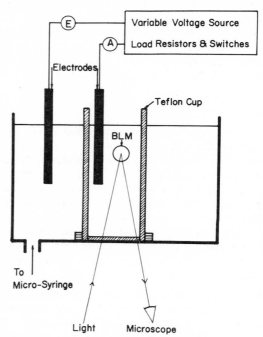

Fig. 20. Arrangement of apparatus used for studying electrical properties. The inner chamber is made of a 10-ml Teflon beaker.

The current-limiting resistance R_S must have a value at least 10-fold that of the BLM resistance R_m, i.e., a value of ca. 1×10^6 to $1 \times 10^{10} \Omega$. True variable resistors in this range are not available, and R_S must be a switched series of resistors (such as the glass-sealed high megohm resistors available from Victoreen Instrument Co., Cleveland, Ohio).

The voltage-measuring instrument, generally an electrometer millivolt-meter, must have a high input impedance (greater than $1 \times 10^{10} \Omega$) and should provide a low impedance output for a strip chart recorder. Most tube-type pH meters are electrometers with millivolt calibration and a sufficiently high input impedance. "Solid state" pH meters and electrometer voltmeters in some instances use an electrometer tube input and in other cases use a junction-type field effect transistor input. Either type is satisfactory, again provided the input impedance is sufficiently high (excellent instruments are the Models 610B, 602, and 610C electrometers, Keithley Instruments, Cleveland, Ohio). The current of BLM systems may be measured with a picoammeter (Model 416 or 417, Keithley Instruments).

The electrical properties of BLM may be determined by ac methods using an equivalent bridge circuit, as shown in Fig. 21. Platinum electrodes coated with platinum black are generally used. The apparatus should be sufficiently sensitive to detect bridge balance with a signal voltage of no more than ca. 20 mV peak to peak. Detection may be achieved either with a wide-band cathode ray oscilloscope or with a tuned amplifier and null detector.

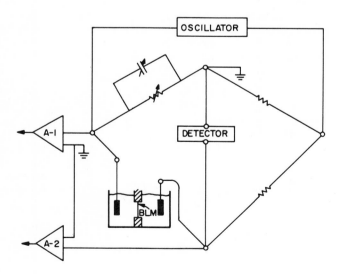

Fig. 21. Schematic diagram showing an experimental setup used to determine the BLM resistance and capacitance. A-1 and A-2 are high input impedance amplifiers ($10^{10} \Omega$). See text for other details.

Commercial instruments suitable for these purposes are the Universal Bridge B221 (Wayne Kerr Laboratories, Ltd.) and the Radio Frequency Bridge B601 (15 kc/sec to 5 Mc/sec). The Model 1210-C and Model 1232-A tuned amplifier and null detector (General Radio Corp.) are compatible for use with the above bridges.

Simultaneous measurement of transmembrane voltage and transmembrane current under varying signal conditions, or under conditions in which rapid changes have been induced in membrane properties, may be made with the equivalent electrical circuit shown in Fig. 19. Four electrodes are utilized in this system, two current electrodes and two voltage electrodes. A current electrode on one side of the BLM is connected through the limiting resistor to the signal source(s) and thence to ground. The current electrode on the other side of the BLM is connected to the summing point of an operational amplifier whose output is used to drive the lower beam of a dual-beam cathode ray oscilloscope and (fed back through an appropriate high resistance) to maintain that side of the BLM system at essentially ground potential. The voltage electrodes are connected differentially through high input impedance preamplifiers to the upper beam of the oscilloscope.

Monitoring of membrane formation, thinning, and rupture may be achieved by recording a function of membrane impedance using the equivalent circuit shown in Fig. 19. The capacitatively coupled ac amplifier, rectifier, and strip chart recorder may be connected in a similar manner to the output of the electrometer preamplifier(s) of Fig. 19. Direct current offset voltages resulting from applied dc signals, electrode polarization, or transmembrane electrochemical potentials do not affect this record, because of the RC coupling. By appropriately increasing the recorder time constant, brief alterations (less than 2 sec duration) of BLM impedance (resistance) are not recorded either, and a record related primarily to membrane thickness is

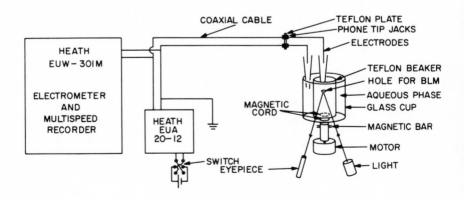

Fig. 22. Block diagram of an inexpensive setup for the measurement of BLM electrical parameters (see Appendix D for a listing of parts).

obtained. With a very short recorder time constant and fast recorder response, even brief alterations in membrane impedance may be observed.

5. Parts List for an Inexpensive Setup

For beginner and student purposes, and for those with modest resources, a listing of inexpensive apparatus and components is given in Appendix D, along with the names of the suppliers. These suggestions are provided for the purpose of aiding the investigator to begin his experiments. The recommended setup (Fig. 22), although inexpensive, is fully adequate for the new investigator and has been used in the authors' laboratory for the instruction of BLM techniques. With this setup, the investigator can measure most of the basic electrical properties of BLM.

C. BLM Resistance

1. Direct Current Method

The electrical resistance is usually measured by impressing a small voltage across the BLM via a pair of reversible electrodes [Eq. (26)]. A variable dc voltage source can be constructed from a small Hg dry cell and a precision potentiometer (10-turn Helipot). Figure 19 shows the simple circuit diagram. Normally the voltage drop across the BLM is measured, and from a knowledge of the applied voltage the BLM resistance is readily calculated according to Ohm's law:

$$R_M = [E_m/(E_i - E_m)] R_s \qquad (27)$$

where R_s is the series resistance, E_i is the calibrated input voltage, and E_m is the voltage appearing across the BLM. Current/voltage (IV) curves can be calculated from the BLM potential responses as a function of applied voltage. For an unmodified BLM a linear relationship is generally found between membrane current and voltage, when the membrane potential is less than 50 mV. When an electrometer together with a picoammeter is used in the measuring circuit, both E_m and I_m may be determined directly and simultaneously.

The experimental transmembrane resistance R_M (in Ω) is converted to the normalized transmembrane resistance R_m (in Ω-cm^2) by multiplying R_M by the measured BLM area A. If A is expressed in square centimeters, then the working equation is:

$$R_m = R_M \cdot A \qquad (28)$$

The normalized value R_m is the value generally reported (Table V).

TABLE V

Electrical Resistance of Bimolecular Lipid Membranes (BLM)[a]

Lipid	Solvent	Aqueous Phase	Resistance
Brain lipids + α-tocopherol	$CHCl_3 : CH_3OH$	0.1 N NaCl	10^7-10^8
Brain lipids (+ EIM)	$CHCl_3 : CH_3OH$	0.1 N NaCl	10^4
Lecithin (egg)	n-Decane	Various	10^8-10^{10}
Oxidized cholesterol	Octane	0.1 N NaCl	10^8
Cholesterol + dodecyl acid phosphate	Dodecane	0.1 N NaCl	10^7-10^8
Cholesterol + dioctadecyl phosphite	Dodecane	0.1 N NaCl	10^8-10^9
Cholesterol	Dodecane	0.001 N NaCl + 0.008% HDTAB[b]	10^7
		0.1 N NaCl + 0.008% HDTAB[b]	10^3-10^4
Chloroplast extracts	Octane + butanol	Various NaCl	10^5-10^6

[a] For further details, see Ref. [4].

[b] Hexadecyltrimethylammonium bromide.

The BLM resistivity (specific resistance, in Ω-cm) is calculated from R_m (or R_M and A_m) using the equation

$$\rho_m = R_m/t_m \tag{29}$$

where t_m is the BLM thickness (often assumed to be 50 Å or 5×10^{-7} cm). Note that the correct dimensions for R_m and ρ_m are Ω-cm^2 and Ω-cm, respectively, and not Ω/cm^2 or Ω/cm.

Unmodified BLM usually exhibit resistances R_m around 10^6-10^8 Ω-cm^2, with corresponding resistivities ρ_m calculated as 2×10^{12} to 2×10^{14} Ω-cm for an assumed membrane thickness of 50 Å. This resistivity range corresponds to that of most bulk hydrocarbons.

BLM resistance is commonly irreproducible from one membrane to another, although the resistance of a given membrane is usually relatively constant until a short time before the membrane ruptures. Relative changes in the resistance of a given BLM, due to addition of other substances or treatment in the system, may therefore be accurately determined, although the absolute values of resistance and its alteration will vary over an order of magnitude with different BLM.

The degree of irreproducibility of R_m is probably due to ion-conductive leakage pathways between the aperture and the Plateau-Gibbs border and may be decreased by forming membranes by the injection method and by rejecting for study any membranes which exhibit visual "crystals" of lipid at the BLM periphery.

A more sophisticated study of BLM resistance may be made by simultaneously plotting E_m on one rectangular coordinate axis and I_m on the other, using either an X-Y recorder or an oscilloscope in X-Y mode, as I_m is slowly increased. The resulting I vs. V plots (Fig. 23) demonstrate the ohmic nature of BLM at potential levels below 30 to 60 mV, with a gradual decrease in apparent resistance at higher potentials until membrane breakdown occurs (see Sec. IV. C. 2).

2. Alternating Current Method

The experimental measurement of the complex transmembrane impedance, $\overline{Z}_m$, may be accomplished either by a comparison procedure using an impedance bridge, or by sine wave analysis. Only the former method will be discussed in detail; the reader is referred to other works for the method of sine wave analysis.

In the comparison method using the bridge circuit of Fig. 21, if R_1 is equal to R_2 and the bridge is balanced (at null, no detectable voltage differences between points b and c of the circuit), the unknown impedance $\overline{Z}_m$ must be equal to the impedance $\overline{Z}_v$ of the variable resistance R_v and capacitance C_v. Although many combinations of R_v and C_v would yield an impedance numerically equal to $\overline{Z}_m$, only one combination will also yield the same phase angle as the unknown. The voltage signal across the bridge will be zero at all times only when the numerical values of $\overline{Z}_v$ and $\overline{Z}_m$ are equal, when the amplitude of the two sine wave voltage signals across them are equal, and when there is no difference in phase angle between these voltages.

The complex impedance $\overline{Z}_m$ is analyzed by the equation:

$$Z_m = (R_2/R_1) \left[R_v - (j/\omega C_v) \right] \tag{30}$$

where ω is the angular velocity (expressed in rad-sec^{-1}, equal to $2\pi f$ where f is the sine wave frequency in cycles/sec), j is $\sqrt{-1}$, and other symbols are

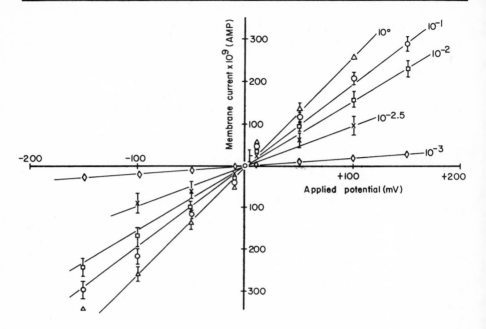

Fig. 23. Linear plots of the current/voltage characteristics of chloroplast BLM in various NaCl solutions. The ordinate is plotted as membrane
current per square centimeter of area.

as defined above. The real and imaginary components of Eq. (30) may be
separated into one independent equation containing all real terms (resistive)
and a second equation containing all imaginary terms (reactive), as given by

$$R = R_2 R_V / R_1 \tag{31}$$

$$X = -1(R_2 / R_1) \tag{32}$$

Thus, the real impedance Z_m is

$$Z_m = (R_2 / R_1)\left[R_V^2 + (1/\omega C_V)^2 \right]^{\frac{1}{2}} \tag{33}$$

and the phase angle ϕ is

$$\phi = \frac{\cos^{-1} R_V}{\left[R_V^2 + (1/\omega C_V)^2 \right]^{\frac{1}{2}}} \tag{34}$$

Since the BLM impedance Z_m may be considered to be due to a combination of
R_m and C_m, then

$$R_m - (j/\omega C_m) = (R_2/R_1) \, [\, R_v - (j/\omega C_v) \,] \qquad (35)$$

Separating this provides the working equations:

$$R_m = R_2 R_v/R_1 \qquad (36)$$

$$C_m = R_2 C_v/R_1 \qquad (37)$$

The conditions of balance given by Eqs. (36) and (37) for such a series-capacitance bridge contain no frequency terms. This type of measurement may thus be made over a wide range of frequencies and is, at least theoretically, independent of wave form.

In general, it is simpler to use dc methods in assessing the BLM impedance; they cause a transient response, however, owing to the BLM capacitance. The advantage of the ac methods is that they permit observation of fast changes when investigating interactions between the BLM and its modifiers. In using ac methods, the applied frequency must be carefully selected so that the effect of capacitative shunting is minimized. Furthermore, the chosen frequency must not be close to the line frequency, i.e., 50, 60, or 400 Hz. Some preliminary measurements indicate that a frequency around 350 Hz is a suitable one to use. It may also be pointed out that with ac methods electrode polarization is less likely to occur.

D. Dielectric Breakdown Voltage

The breakdown voltage at which BLM rupture occurs may range from 80 to 500 mV or more, depending on the duration of the applied voltage, the chemical composition of the BLM and the aqueous phase, and the past history of the membrane. A typical value of E_m for breakdown of a phospholipid BLM in 0.1 N NaCl, as determined from the I vs. V plots described in Sec. IV.B.4 is 150 mV. Experimentally, the applied voltage, E_i, is gradually increased until the BLM ruptures. The dielectric breakdown voltage, V_B (in V-cm^{-1}), may be calculated from the measured transmembrane voltage, E_m, at which breakdown occurs and the dielectric thickness, t_ε, by the equation:

$$V_B = E_m/t_\varepsilon \qquad (38)$$

For the 150-mV breakdown voltage given in the example above, and assuming a dielectric thickness of 50 Å, the dielectric breakdown voltage would be 3×10^5 V-cm^{-1}. This value compares with some of the best insulators known, such as unfilled flexible polyethylene ($V_B = 4 \times 10^5$ V-cm^{-1}) or solid paraffin ($V_B = 1 \times 10^5$ V-cm^{-1}).

E. BLM Capacitance

Next to the electrical resistance, the capacitance of the BLM is the second most measured parameter. Experimental results have shown that the capacitance of the unmodified BLM is usually independent of frequency from about 10^{-3} to 5×10^{-6} cycles/sec [61].

1. Alternating Current Bridge Method

Accurate measurements of BLM capacitance can be made by using the familiar Wheatstone bridge (see Sec. IV. B for apparatus). Platinum electrodes preferably coated with platinum black should be used. An oscilloscope with a preamplifier in the range 50 Hz to 50 kHz is generally used to detect the bridge balance. At the null point, the impedance of the BLM can be determined in terms of an RC equivalent circuit. A brief consideration of the theoretical aspects of bridge measurements is given in the following paragraphs.

In electrical terms, a capacitor (condenser) consists of two good conductors separated by a poor conductor. A BLM separating two aqueous phases meets this requirement since a typical BLM has a resistance many orders of magnitude higher than the adjacent aqueous solutions. Therefore, the BLM is considered as a parallel-plate condenser in parallel with its resistance. This assumption has permitted an evaluation of the BLM thickness as discussed in Sec. III. B. In Fig. 21 the equivalent parallel combination of capacitance and conductance of the whole assembly (membrane chambers, electrodes, lead wires, etc.) is illustrated. It should be noted that the conductance, G ($=1/R$), rather than resistance, R, is used, since parallel conductances are simply additive. In general, the stray capacitance is less than 100 pF and the corresponding conductance shunting the BLM is less than 10^{-14} mho. These quantities are therefore usually neglected in the calculation. The total capacitance of the assembly in the absence of a BLM is given by

$$C_{w/o} = C_s + C_o + (G_s/j\omega) \tag{39}$$

where C_s and C_o are the capacitances for solution and cell chamber, respectively, G_s is the conductance for the aqueous phase, $j = \sqrt{-1}$, and $\omega = 2\pi f$. In the presence of a BLM, the capacitance is given by

$$C_{w/m} = C_s + C_a + \frac{C_\ell - C_a}{1 + \omega^2 \lambda^2} - j \left[\frac{(C_\ell - C_a)\omega\lambda}{1 + \omega^2 \lambda^2} + \frac{G_\ell}{\omega} \right] \tag{40}$$

in which C_a is the capacitance near the aperture and is given by

$$C_a = C_s C_m / (C_s + C_m) \tag{41}$$

The other quantities are:

$$C_\ell = (C_s G_m{}^2 + G_s{}^2 C_s)/(G_s + G_m)^2 \tag{42}$$

$$G_\ell = G_s G_m/(G_s + G_m) \tag{43}$$

$$\lambda = (C_s + C_m)/(G_s + G_m) \tag{44}$$

where G_m is the BLM conductance and C_m is the BLM capacitance. It should be mentioned that the capacitance and conductance of the P-G border around the BLM are negligible compared with those of the BLM.

Since $G_s \gg G_m$ and $\omega \sim 10^4$, Eq. (40) may be simplified to

$$C_{w/m} = C_o + C_m - \frac{1}{j} \left[\frac{(C_m - C_a)\omega\lambda}{1 + \omega^2\lambda^2} + \frac{G_m}{\omega} \right] \tag{45}$$

In the bridge method the measured quantity is usually given in terms of the parallel components and is given by

$$C_{w/m} = C_t + G_{t/jw} \tag{46}$$

But $C_t = C_o + C_m$ and $C_m \gg C_o$, therefore

$$C_t \simeq C_m \tag{47}$$

When the total capacitance is plotted as a function of its conductance divided by the frequency, a typical semicircular curve is obtained [31]. This plot can then be utilized to obtain the limiting value of the capacitance at low frequency. The low frequency capacitance is insensitive to the concentration of 1-1 electrolyte used (e.g., NaCl), which suggests that only the hydrocarbon portion of the BLM is involved in the dispersion. At high frequencies the conductance of the system tends to that of the aqueous solution bathing the membrane, as would be expected.

2. Direct Current Transient Method

The capacitance of a BLM may also be measured by a dc transient (charge leakage) method using a standard high resistor. A typical circuit connection using a Keithley 610 electrometer is shown in Fig. 16. This method simply involves applying a small dc voltage across the BLM. After reaching a steady state, as indicated on a monitoring device (strip chart recorder or oscilloscope), the switch is opened and the decay of the membrane voltage (E_m) with time is recorded. A typical charging and discharging curve is shown in Fig. 24. The BLM capacitance is calculated from the following equation:

$$C_m = \frac{t}{R_p \ln (E_o/E_t)} \tag{48}$$

where t is the time (in seconds), E_o is the steady-state reading at t = 0, E_t is the voltage reading at time t, and R_p is the leakage resistance which is given by

$$R_p = R_m R_s/(R_m + R_s) \tag{49}$$

The dc transient method is straightforward and simple to apply and is generally preferred by many investigators.

The capacitance of the BLM is usually expressed in $\mu F/cm^2$. In Table VI some representative values of C_m for a variety of BLM are given.

3. Discussion

The BLM capacitance is of special interest, as can be seen in Eq. (23). It is related to the membrane thickness, the dielectric constant of the membrane, and through classical electrochemistry to the electrical double layers at the biface. It is therefore informative to discuss the BLM capacitance in terms of electrolyte concentration, membrane charge density, and the effect of applied potential. Some aspects of the BLM capacitance and its associated electrical double layers have been considered [62].

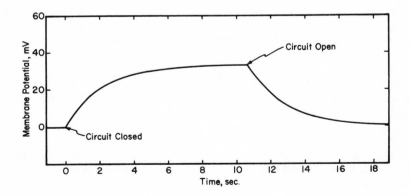

Fig. 24. A typical charging and discharging curve for a BLM formed from a mixture of cholesterol and HDTAB in 0.001 NaCl solution.

TABLE VI

Capacitances of Bimolecular Lipid Membranes (BLM) [a]

Lipid	Solvent	Aqueous phase	Capacitance, $\mu F/cm^2$
Brain lipids + α-tocopherol	$CHCl_3 : CH_3OH$	0.1 N NaCl	0.7-1.3
Lecithin (egg)	n-Decane	Various 0.1 N NaCl	0.38 0.35
Lecithin (egg) + cholesterol	n-Decane	0.1 N NaCl	0.28-0.56 [b]
Oxidized cholesterol	n-Octane	0.1 N NaCl	0.57
Cholesterol	Dodecane	0.001N NaCl + 0.008% HDTAB [c]	0.79
Cholesterol + dodecyl acid phosphate	Dodecane	0.1 N NaCl	0.69
Cholesterol + dioctadecyl phosphite	Dodecane	0.1 N NaCl	0.74

[a] For further details, see Ref. [4].

[b] Value depends on proportion of each component.

[c] Hexadecyltrimethylammonium bromide.

Equation (23), used for the calculation of the membrane thickness (Sec. III.B), is of basic importance and will be derived [63]. The usual diagram of a BLM separating two aqueous phases is shown in Fig. 25. As a result of the orientation of amphiphilic molecules at the biface, the density and sign of the interfacial potential would depend on the relative numbers and nature of charged groups at the biface. Depending upon the ionic strength of the aqueous solution, the electrostatic field will extend out into the bulk phases at varying distances and will decrease to zero at a short distance from the biface.

The equation for the BLM capacitance may be derived from a consideration of the electrical double layers at the biface. Figure 25 shows a BLM of a given thickness separating two aqueous solutions. It is assumed that each side of the biface has a uniform charge density, σ_0, in contact with a

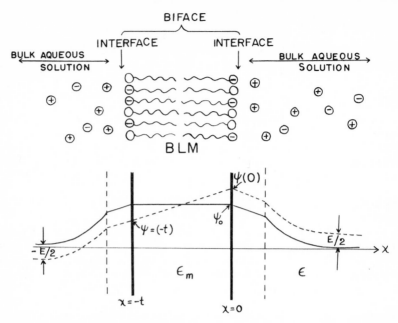

Fig. 25. Schematic diagram illustrating a BLM separating two aqueous phases. Electrical potential ψ as a function of the distance x in the absence of an applied voltage is shown by the solid line, whereas the dotted line indicates the potential profile in the presence of an applied field E. ϵ_m and ϵ are the dielectric constant of the BLM and the aqueous phase, respectively. The potential on the left-hand side is lowered by E/2 and is raised by E/2 on the right-hand side, as shown.

solution of a uniunivalent electrolyte ($z = 1$) of concentration C. In the absence of an applied voltage, the potential profile as a function of distance from the biface is symmetrical (Fig. 25 A). During capacitance measurements, an applied voltage, E, is impressed across the BLM. The potential in the bulk solution on one side of the biface changes by E/2 and on the other side by -E/2. It is also assumed that there is no electrolyte present in the BLM. In the presence of the applied voltage E, a space-charge layer appears across the BLM. The potential profile will be altered. The space-charge layer is built up between $x = -t$ and $x = 0$. The electrical potentials at the biface are taken to be $\psi(-t)$ and $\psi(0)$. The potential at any point determines the potential energy of the ion in the electric field and is equal to $ze\psi$. According to the Boltzmann equations, the concentrations of n_+ (cation) and n_- (anion) are given by

$$n_+ = n_0 \, e^{-ze\psi(\chi)/kT} \tag{50}$$

$$n_- = n_0 \, e^{ze\psi(\chi)/kT} \tag{51}$$

where n_0 is the bulk phase concentration. $\psi(\chi)$ is given by

$$\psi(\chi) = (ze/kT) [\psi(\chi) - \psi(\infty)] \tag{52}$$

for $0 \leq \chi \leq \infty$, and

$$\psi(\chi) = (ze/kT) [\psi(\chi) - \psi(-\infty)] \tag{53}$$

for $-\infty < \chi \leq -t$. In the equations above, k is the Boltzmann constant, e is the electronic charge, ψ is the mean electrostatic potential, T is the absolute temperature, and z is the valence of the ion ($z = 1$). The net charge density, ρ, at any point is given by

$$\rho = e(n_+ - n_-) = 2n_0 e \sinh(e\psi/kT) \tag{54}$$

When $E = 0$, each side of the biface has a charge density ρ_0 which is balanced by an equal and opposite diffuse charge, σ_d, in the bulk solution. An additional diffuse charge will be present when a voltage is applied. For the right side of the biface (Fig. 25) this additional diffuse charge density (σ_a) as a result of applied voltage is given by

$$\sigma_a = \int_0^\infty \rho \, dx - \sigma_d \tag{55}$$

Introducing the Poisson equation, which relates the gradient of the electrical potential at a point to the charge density, for a planar BLM,

$$d^2\varphi/dx^2 = -(4\pi\rho/\varepsilon) \tag{56}$$

where ε is the dielectric constant of the solution. In Eq. (56) replacing σ by Eq. (54), one obtains

$$d^2\varphi/dx^2 = (8\pi e^2 n_0/\varepsilon kT) \sinh \varphi = \varkappa^2 \sinh \varphi \tag{57}$$

where $\varkappa$ is the reciprocal length defined in the well-known Debye-Huckel theory [63]. Integration of Eq. (57) gives

$$(d\varphi/dx)^2 = 2\varkappa^2 \cosh \varphi + \text{const} \tag{58}$$

Under the conditions

$$X \to \pm\infty, \quad d\psi/dx = d\varphi/dx \to 0$$

$$\psi(x) \to \psi(\pm\infty) \text{ and } \varphi(x) \to 0$$

Therefore Eq. (58) becomes

$$\frac{d\varphi}{dx} = \frac{e}{kT} \frac{d\psi}{dx} = \pm\varkappa [2(\cosh\varphi - 1)]^{1/2} = \pm 2\varkappa \sinh(\varphi/2) \tag{59}$$

When x = 0, Eq. (59) becomes

$$\left(\frac{d\psi}{dx}\right)_{x=0} = \pm \frac{2\varkappa kT}{e} \sinh \frac{e}{2kT} \left[\psi(0) - \frac{E}{2}\right] \qquad (60)$$

Combining Eq. (60) with Eq. (55) and remembering that

$$\int_0^\infty \rho\ dx = (\varepsilon/4\pi)(d\psi/dx)_{x=0} \qquad (61)$$

under the boundary conditions $x \to \infty$, $d\psi/dx \to 0$; $x = 0$, $d\psi/dx = (d\psi/dx)_{x=0}$, one obtains

$$\sigma_a = \pm \frac{\varkappa ekT}{2\pi e} \sinh\left(\frac{e}{2kT}\right)\left(\psi(0) - \frac{E}{2}\right) - \sigma_d$$

$$= \pm \frac{\varkappa ekT}{2\pi e} \sinh \alpha - \sigma_d \qquad (62)$$

Since the capacitance of the BLM is given by

$$C = \sigma_a/E \qquad (63)$$

therefore

$$C = \pm (\varkappa eT/2\pi eE) - \sinh \alpha - \sigma_d \qquad (64)$$

The exponential terms in Eq. (64) can be expanded, which is simplified to an expression

$$C = \frac{C_m \sinh \alpha}{(2\alpha/\varkappa t)/(\varepsilon_m/\varepsilon) + \sinh \alpha} \qquad (65)$$

in which $C_m = \varepsilon_m/4\pi t$. As has been discussed by Läuger et al. [62], C_m is the geometrical capacitance of the BLM per square centimeter, which is always greater than the measured total capacitance, C. For small values of applied E, $kT/e \gg E$, $\sinh \alpha \sim \alpha$. Equation (65) then becomes

$$C = \frac{C_m}{1 + (2\varepsilon_m/\varkappa t\varepsilon)} \qquad (66)$$

Equation (66) can be rewritten in a more familiar form:

$$\frac{1}{C} = \frac{1}{\varepsilon_m/4\pi t} + \frac{2}{\varepsilon\varkappa/4\pi} \qquad (67)$$

where the first term on the right-hand side is due to the BLM capacitance and the second term is attributed to the capacitance of the electrical double layers.   Further, it is known that for a 1-1 electrolyte, 1/k is about 10 Å at 0.1 M.   Therefore, for practical purpose the measured $C \simeq C_m$ [33].

In general, it has been found that the BLM capacitance is more reproducible from one membrane to the next, whereas BLM resistance is not usually reproducible.   Earlier workers [31] found that the BLM capacitance is independent of frequency and also of the nature and concentration of the electrolyte.   More recent studies have shown that there are significant changes in the capacitance of certain BLM when either the electrolyte concentration or ionic species is altered [33].   The experimental results are presented in Fig. 26.   Contrary to the prediction of the classical theory [63], a decrease in the capacitance is observed with increasing electrolyte concentration.   To explain this discrepancy, it is suggested that the effective area available to hold charges may be involved when the nature and the salt concentration of the bathing solution are varied [33].   One other possibility is that different ionic species interact with the BLM differently and cause a change either in the BLM thickness or in the BLM dielectric constant.   These suggestions should be a fruitful area for further investigation.

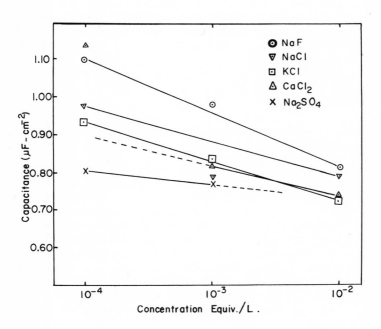

Fig. 26.   Capacitance of BLM formed from cholesterol + HDTAB (0.008%) as a function of various salt concentrations.

F. Transmembrane Potential

The resistance of most unmodified BLM's is ohmic (at $E_m \leqq 50$ mV) when the ionic species and strength are identical in each aqueous compartment, as discussed above (Sec. IV. B). Under such conditions there is no rectification, and there is no residual transmembrane potential in the absence of externally applied current.

A transmembrane potential may be generated, however, when there is a difference in ionic strength or ionic species or both across the membrane. This potential may be measured by the same high impedance input device used in measuring transmembrane voltage due to externally applied current.

The measured potential E_m depends on the ionic species present, the relative ionic mobilities through the membrane, and the ionic activities of the aqueous phases, based on the Nernst relationship for each ion:

$$E_m = \frac{RT}{nF} \ln \frac{a_i}{a_o} \simeq \frac{RT}{nF} \ln \frac{C_i}{C_o} \tag{68}$$

where a_i and a_o are the activity of ion in the inner and outer chamber, respectively. More frequently, the actual concentrations, C_i and C_o, are used, assuming that the activity coefficients are both unity. Equation (68) implies that the BLM is ideally more permeable to one ionic species than to the other. The more general equation, taking into account the transference number of both ions (e.g., 1-1 electrolyte), is

$$E_m = \frac{\mu_+ - \mu_-}{\mu_+ + \mu_-} \; \frac{RT}{F} \; \ln \frac{C_i}{C_o} \tag{69}$$

where μ_+ and μ_- are the mobilities of respective ions. Since $\mu_+/(\mu_+ + \mu_-)$ is equal to the transference number of the cation, i.e., to t_+, Eq. (69) becomes

$$E_m = (2t_+ - 1) \; \frac{RT}{F} \; \ln \frac{C_i}{C_o} \tag{70}$$

Therefore, Eq. (70) can be used to evaluate the transference numbers from a knowledge of the concentrations and experimentally measured E_m. For a certain type of BLM, it has been found that the transference numbers are independent of temperature, the hydrocarbon solvent, and the concentration of the bathing solutions [38]. For most common 1-1 electrolytes, the observed E_m is only a few millivolts for a 10-fold concentration difference. However, in one case using KI-I_2 [64], the measured voltage almost reaches the value predicted by Eq. (68).

The BLM may be considered similar to an ion exchange membrane in which there exists an equilibrium of the type

$$\text{BLM} - C_1^+ + C_2^+ \rightleftarrows \text{BLM} - C_2^+ + C_1^+ \tag{71}$$

where C_1^+ and C_2^+ are exchanging cations, assuming a negatively charged BLM. The selectivity coefficient, K, is defined empirically as

$$K = \frac{(\text{BLM-}C_2^+)^n \,(C_1^+)}{(\text{BLM-}C_1^+)^n \,(C_2^+)}$$

where $n \geq 1$. It has been shown that by assuming a fixed number of exchanging sites in (or on) the BLM, an equation relating the effect of salt concentration in the two chambers and the observed E_m can be derived [65]. The equation, given below, is useful in correlating the ion selectivity of the BLM:

$$E_m = E' + (RT/F) \ln [\,(C_1^+)^{1/n} + (K \cdot C_2^+)^{1/n}\,]^n \tag{72}$$

in which E' is a cell constant. The selectivity values for the five common alkali cations, calculated with the aid of Eq. (72), are given in Table VII.

V. MEASUREMENT OF PERMEABILITY PROPERTIES

Mass transport of substances across BLM is of special interest with regard to several major problems in biological systems, such as active transport, ion selectivity, and drug interaction. From a molecular and physicochemical point of view, the simple BLM system provides a unique opportunity for obtaining basic information and testing working hypotheses. In this section the principles and techniques are presented for three methods of measuring BLM permeability to (respectively) water, radioisotope-labeled substances, and ions. The methods are general in nature, and with minor modifications can be used for measurement of BLM permeability to a variety of solutes.

A. Basic Principles

If it is assumed that permeation of substances through BLM is a process essentially similar to diffusion in liquid systems, the same laws can be applied. Diffusion is the migration of a substance driven by a chemical potential (activity) gradient. The rate of transfer of a substance from a region of greater activity to a region of lesser activity is related to the driving gradient by Fick's first law of diffusion:

$$dn/dt = DA \; da/dx \tag{73}$$

TABLE VII

Ionic Selectivity Coefficients, K_{Li}^{C}, of Bimolecular Lipid Membranes (BLM)[a]

| Cation (C^+) | Brain lipids | BLM generated from: | |
		Sphingomyelin + valinomycin	Lecithin + valinomycin
Li^+	1.0	1.0	1.0
H^+	---	---	135
Na^+	1.2	1.8	0.8
K^+	2.0	11	10
Rb^+	2.2	13	25
Cs^+	2.5	8.8	6.3

[a] Data from refs. [80-82].

where dn/dt is the rate of transfer of n crossing area A, and da/dx is the driving gradient of activity. D is a proportionality constant, called the diffusion coefficient or diffusivity. In practice another factor, called the permeability coefficient (P), is used, given by

$$J = dm/A \, dt = -P \, da \qquad (74)$$

in which J is defined as the net unidirectional flow of mass per unit time and per unit area. The flux may be determined directly as the transfer of solvent or as the transfer of solute, and indirectly as a function of the transfer of ions; hence the three methods discussed below.

B. Osmotic Flux Method

In the case of a membrane which is effectively impermeable to solute but permeable to solvent, an osmotic gradient will parallel the existence of any concentration gradient across the membrane. For a BLM separating two aqueous phases which contain different concentrations of an effectively impermeable solute such as glucose or NaCl, the osmotic pressure gradient will cause a transmembrane volume flux of water, J_w, given by

$$J_w = P_o \, \Delta \Pi \qquad (75)$$

where P_o is the osmotic permeability coefficient. $\Delta \Pi$ is the osmotic pressure gradient given by

$$\Delta \Pi = \nu RT (\phi_2 c_2 - \phi_1 c_1) \tag{76}$$

where ν is the number of particles of the solute on dissociation, R and T are the gas constant and and absolute temperature, respectively, ϕ is the standard osmotic coefficient, c is the solute concentration, and the subscripts 1 and 2 identify the two aqueous solutions.

The principle of the osmotic flux method is that the transmembrane volume flux of water will create a hydrostatic pressure gradient which will bulge the BLM. The membrane is maintained in a planar state by adjusting the volume of one aqueous phase, and the volume flux is determined directly from the volume adjustment made in a given time interval.

Experimentally, a cell of the type shown in Fig. 27 is often used for the osmotic flux method [69]. The cell consists of a closed inner chamber made from a 10-ml Teflon beaker. The volume of the inner compartment (containing aqueous solution 1) can be adjusted by manipulation of a precision micrometer syringe (0.20-ml capacity, Roger Gilmont Instruments, Great Neck, N.Y.) or a precision microliter syringe (0.25-ml capacity, Cole-Palmer Instrument and Equipment Co., Chicago, Ill.). After the membrane has become black, a known quantity of a concentrated solution, e.g., NaCl, is added to the outer compartment (containing aqueous solution 2) and thoroughly mixed. The osmotic pressure gradient will cause the efflux of water from solution 1 across the BLM into solution 2. This efflux is manifested by the inward bulging of the BLM. The BLM is observed visually and a planar membrane or constant pattern of reflected light is maintained by manipulating the micrometer syringe.

The volume increments ΔV are read directly from the micrometer syringe at intervals Δt, and P_o is calculated from

$$P_o = \frac{\Delta V / \Delta t}{A \nu RT (\phi_2 c_2 - \phi_1 c_1)} \tag{77}$$

where A is the measured BLM area. Experimentally determined values of P are often expressed in μ/sec or in $\mu^3/\mu^2/min/atm$ ($= \mu/min/atm$), where $\mu = 10^{-4} cm$. The conversion factor between the two units is

$$P_{(\mu/sec)} = (0.925RT)P_{(\mu/min/atm)} \tag{78}$$

where R is the gas constant and T is the absolute temperature.

C. Diffusion Flux Method

In the case of a membrane whose finite permeability to a substance is to be determined by direct measurement of the permeant substance, Fick's law

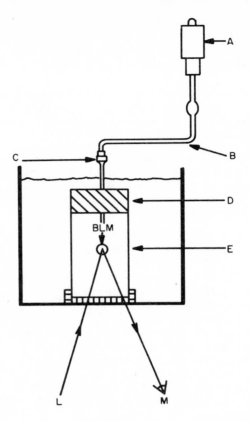

Fig. 27. Apparatus for measuring osmotic water flux across BLM. A-digital pipet, B-polyethylene or Teflon tubing, C-syringe needle, D-stopper, E-Teflon beaker, L-light source, M-binocular microscope.

may be rewritten:

$$J = dm/dt = P_d A(c_2 - c_1) \tag{79}$$

where J is the molar diffusion flux, the activity coefficients are assumed to be unity, and a different proportionality constant (called the diffusion permeability coefficient, P_d) is introduced because the effective membrane thickness and concentration gradient are generally unknown.

The diffusion permeability coefficient of the membrane for a given substance is obtained by rewriting Eq. (79) in the explicit terms of the experimental parameters:

$$P_d = V_1 \Delta c_1 / A(c_2 - c_1) \Delta t \tag{80}$$

where V_1 is the volume of aqueous phase 1 in which a concentration change Δc_1 has occurred at interval Δt, and c_1 and c_2 are the average concentration in aqueous phases 1 and 2 during Δt. If the initial value of c_1 is zero, Δc is much less than c_2, and c_2 is essentially constant over Δt; then

$$P_d = V_1 \Delta c_1 / A c_2 \Delta t \qquad \qquad (81)$$

Experimentally, a known amount of a substance for which BLM permeability is to be determined is added to aqueous phase 2 after the membrane has become black. Sequential aliquots are removed from both aqueous phases at regular time intervals and the concentration of the substance is determined in each aliquot. Because of the extremely low concentrations in Δc_1, the two most useful methods have been liquid scintillation spectrometry for β-radioisotope-labeled substances and direct fluorescence spectrophotometry for fluorescent substances [29].

In experiments to measure BLM permeability to β-radioisotope-labeled substances such as tritiated water (THO), a cell of the type shown in Fig. 28 is used, containing a 5-ml Teflon cup which can be sealed gastight. A BLM is formed on the unsealed cup and a known volume of known specific activity THO is pipetted into the cup (aqueous phase 2) through an opening in the cover. The cup is then sealed (the outer compartment can be simply a large beaker). Stirring for both compartments can be accomplished using magnetic fleas (Fig. 19). Generally, 0.05-ml samples for counting are taken from the outer compartment at 5-min intervals when 100 μc of THO are used. After a sufficient lapse of time (approximately 1 hr or so, depending upon the stability of the membrane) a 0.05-ml sample is taken from the cup, and the radioactivity of all samples is measured. The specific activity of the inner compartment (cup) is plotted vs. time, and P_d is calculated from the slope of the line J^* and the working equation

$$P_d = \frac{J^*}{A(c_2{}^* - c_1{}^*)} \simeq \frac{J^*}{A c_2{}^*} \qquad \qquad (82)$$

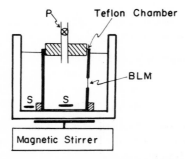

Fig. 28. Schematic diagram of cell used for measuring tritiated water flux across BLM. P-port for introducing THO, S-stirring bars.

where c_1* and c_2* are the average specific activities of the substance in aqueous phases 1 and 2.

D. Electrical Potential Method

The relative permeability of BLM to small ions can be determined by measurement of the transmembrane potential difference generated by a difference in ionic strength or ionic species or both across the membrane, as mentioned in Sec. IV. E. There are three types of determinations which can be made by this method: (1) calculation of the relative permeability of the BLM for cations vs. anions (or a specific cation vs. its counteranion), (2) calculation of the relative permeability of one cation to another cation or of one anion to another anion, and (3) calculation of permeability coefficients.

The simplest determination is that of relative permeability of the BLM for a specific monovalent cation vs. its monovalent counteranion when both aqueous phases are a simple solution of a single uniunivalent electrolyte. In the absence of externally applied current, the transference number t_i for each ion can be determined from the transmembrane potential difference E_m using

$$\sum_{i=1}^{n} t_i = 1 \tag{83}$$

and

$$E_m = \sum_{i=1}^{n} t_i E_i \tag{84}$$

where there are n ions of the i-th type present in the aqueous media, t_i is the ionic transference number, and E_i is the equilibrium potential for the i th ion.

The ionic transference numbers are defined as

$$t_i = G_i/G_m \tag{85}$$

where G_i is the single ion conductance of the i-th ion and G_m is the total transmembrane conductance (equal to R_m^{-1}).

The equilibrium potentials, taking the polarity of side 1 relative to side 2, are calculated from

$$E_i = (RT/Z_i F) \ln (C_{i2} f_{i2}/C_{i1} f_{i1}) \tag{86}$$

where R is the gas constant, T is the absolute temperature, Z is the valence, F is Faraday's constant, and c and f refer to the molal ionic concentrations and ionic activity coefficients, respectively, of aqueous phases 1 and 2.

It is apparent that in the case of uniunivalent electrolytes, if the BLM exhibits no selection for cation vs. anion, the equilibrium potentials will be equal in magnitude but opposite in sign for any ionic concentration gradient, and the measured transmembrane voltage will be zero. If the BLM is selectively more permeable to ions of one type, a concentration–dependent potential will develop. In the extreme case in which the BLM is permeable to ions of one charge and effectively impermeable to ions of the opposite charge, the electrochemical potential will approach the equilibrium potential for the permeable species (see Sec. IV. E).

An electrochemical potential may also be developed across a selective BLM which separates different cations (or anions) with a common counterion, e.g. , 0.01 M KCl vs. 0.01 M LiCl. Such potentials are often called bi-ionic potentials and may also be analyzed using Eqs. (83) – (86). The relative permeability of one ion to another through the BLM is expressed by the ionic selectivity ratio constant K in the equation

$$K_{ix/iy} = t_{ix}/t_{iy} \tag{87}$$

where ix and iy are different ions of the same charge, t_{ix} and t_{iy} are their respective transference numbers calculated from bi-ionic potential measurements, and the ratio constant K is usually referred to the same ion (e.g. , Li^+) for comparison purposes.

The ionic permeability coefficient P_i for each ion may be determined using the calculated equilibrium potential E_i for the ion and the calculated ionic transference number t_i, if the total ionic membrane conductance G_m and the membrane area A are determined. The equation for P_i with uniunivalent electrolytes is

$$P_i = \frac{E_i t_i}{FR'_m (c_{2i} - c_{1i})} \tag{88}$$

Experimentally, the BLM is formed with both aqueous phases at identical concentration (e.g. , 0.01 M KCl), and the electrolyte concentration in one compartment is then altered either by addition of more concentrated electrolyte or by exchange of the aqueous media (e.g. , to obtain 0.10 M KCl on one side). The transmembrane voltage is measured (see Sec. IV. C) first in the absence of externally applied current in order to determine the potential difference E_m due to the ionic concentration gradient, and then in the presence of an externally applied constant current in order to determine the total transmembrane conductance $G_m = 1_m/(V_m - E_m)$. The BLM area A is also measured. The ionic transference numbers and ionic selectivity are calculated using Eqs. (83) – (87). The ionic permeability constant P_i is calculated from Eq. (88). Table VIII gives representative values of t_i for several ions with unmodified BLM.

VI. MEASUREMENT OF BIFACIAL TENSION AND THERMODYNAMIC PROPERTIES

A basic knowledge of the nature of solution–BLM–solution biface is required in order to understand the formation characteristics and stability of

TABLE VIII

Transference Numbers in BLM Under Various Conditions[a]

BLM from:	Solvent	Electrolyte	T_{Na^+}	t_{K^+}	t_{Cl^-}
Egg lecithin	Decane	KCl	---	0.59	0.41
		NaCl	0.55	---	0.45
Red blood cell lipids (high K^+ content)	Decane	NaCl	0.82	---	0.18
		KCl	---	0.84	0.16
Red blood cell lipids (low K^+ content)	Decane	NaCl	0.78	---	0.22
		KCl	---	0.80	0.20
Red blood cell lipids (high K^+ content)	Decane	NaCl + sucrose	0.82	---	0.18
		KCl + sucrose	---	0.82	0.18
	Heptane	KCl	---	0.88	0.12
	Octane	NaCl	0.90	---	0.10

[a] From Ref. [38]

the membrane. In addition, the role played by the Plateau-Gibbs border and its adhesion to the BLM support are interesting and fruitful topics for investigation. In general, the stability of BLM is a complex problem because one has to take into consideration such factors as BLM area, properties of interfacial active material and solvent, as well as the nature of the interacting forces. At present, the most well-known mechanical property which is accessible to direct experimental measurement is the bifacial tension of the membrane. In this section we shall consider a simple technique for the measurement of bifacial tension of the BLM and the calculation of the thermodynamic properties of the biface. The applicability of Gibbs' adsorption equation will also be discussed.

A. Theory

The elementary theory of the maximum bubble pressure method assumes that the maximum pressure is reached when the membrane is just hemispherical [50]. It is implicit in the theory that an interfacial tension (γ_1) exists in every thin film separating two phases (liquid, solid, or gaseous). Further, the interfacial tension is assumed to be the same at every point in a given

film. Considering a planar BLM formed on a hole in a Teflon partition sep-
arating two aqueous phases (Fig. 29), let the radius of the hole be R (cm).
By raising the solution on one side of the membrane, a hydrostatic pressure,
P, is created across the BLM. The bifacial tension is designated by γ_{BLM}.
In Fig. 29(B) the radius of the BLM increases from R to (R + dR). It can be
shown that at equilibrium the work done in changing the radius of BLM bubble,
R, to (R + dR) is equal to $P(4\pi R^2 dR)$. Thus

$$16\pi\gamma_i R dR = P(4\pi R^2 dR) \tag{89}$$

or

$$\gamma_i = Pd/8 \tag{90}$$

where d = 2R, the diameter of the BLM bubble. From Eq. (90), P is at a
maximum when the BLM reaches a hemispherical shape whose diameter is
exactly equal to the diameter of the hole in the Teflon support. In certain
instances where the diameter of the BLM bubble may not correspond to the
diameter of the hole, an alternative method is necessary to ascertain the
BLM bubble diameter. It can be readily shown that the radius of curvature of
a BLM is related to the BLM area and the volume of solution used to bulge
the membrane. Since the effects of the force of gravity on the membrane and
of local pressure differences in the membrane are insignificant (this may not
be true if other force fields are present), the radius of curvature may be assumed to be the same everywhere in the BLM. If the BLM area is designated
as A, and the volume of solution used to generate the hydrostatic head is V,

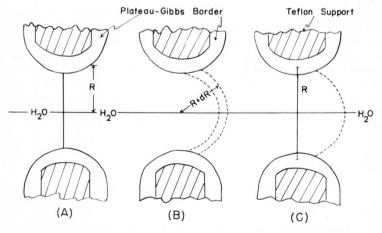

Fig. 29. The bulging out of a BLM under hydrostatic pressure. (A)-
planar configuration, (B)-intermediate configuration, (C)-hemispherical configuration at the maximum pressure difference.

and R_0 is the diameter of the hole in the support, then

$$V = (\pi/6) [(A/\pi) - R_0]^{1/2} [(A/\pi) + 2R_0^2]$$ (91)

and

$$R = (A/2) (\pi A - \pi^2 R_0^2)^{-1/2}$$ (92)

The γ_{BLM} may than be obtained from the slope of a plot of P vs. the reciprocal of R [32].

B. Experimental Techniques

None of the available methods for interfacial tension measurements is easily adaptable to the study of BLM. The so-called interfacial films which have been extensively investigated in the past [50] referred mainly to films at bulk air-water (or oil-water) interfaces. The γ_i of these films were deduced from measurements of surface pressure using either the Langmuir trough or the Wilhelmy hanging plate technique. The BLM system differs from these interfacial films in that a BLM possesses a biface (or two coexisting solution-membrane interfaces). Since it is known that the bifacial tension of a BLM is generally low (less than 6 dyn/cm), the pressure difference that can be imposed across the membrane is correspondingly very small [see Eq. (90)]. Therefore, a very sensitive pressure detection device is essential for the measurement of BLM tension. The apparatus developed for the purpose is shown in Fig. 30 and consists primarily of a very sensitive pressure transducer (Sanborn Model 270 and Pace Model P-90D are suitable). The other major items of the setup include an infusion-withdrawal pump, an amplifier for the transducer, and a strip chart recorder. The temperature of the membrane chamber can be maintained by flowing water from a constant temperature bath through a glass coil placed in the membrane chamber. The temperature should be controlled to within $\pm 0.1°C$. The inner chamber of the membrane cell for BLM support is made of a Teflon sleeve held in place by ground-glass joints. The small hole in the Teflon sleeve may be punched as follows: A simple device is made of two pieces of steel plate welded together with a small clearance. A smooth hole is first drilled through the plates, then fitted with a carefully machined rod of hardened steel. The hole in the Teflon sleeve is punched by ramming the rod through the holes in the plates while the sleeve is placed in the clearance. A smooth hole without frayed edges can be made in this manner.

To form a BLM, a small amount (~ 0.005 ml) of lipid solution is applied via a Teflon capillary attached to a microsyringe (see Sec. II. C). The formation characteristics leading to the black state should be observed under reflected light at 20 to 40X magnification. Other precautions that should be exercised in producing BLM are essentially those previously described. After the membrane has become completely black (except at the P-G border), the

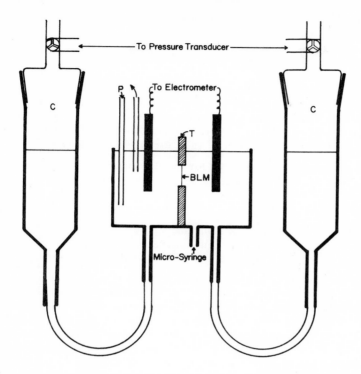

Fig. 30. Schematic diagram of apparatus for measuring bifacial tension of BLM. C-ballast chambers, P-tubings for infusion or withdrawal of aqueous solution, T-Teflon partition (see also Fig. 4). Introduction of electrodes permits the study of the bifacial tension of a BLM as a function of applied voltage.

infusion-withdrawal pump is started. The pressure difference across the BLM is continuously monitored and reaches a maximum when the membrane is hemispherical. The interfacial tension is calculated from this point using Eq. (90).

In case the diameter of the BLM hemisphere does not equal the diameter of the hole in the support, the diameter (or the radius) of the BLM may be evaluated using Eq. (92), which entails the measurement of the BLM area. It was shown in Sec. III.B that the area of a BLM can be estimated from a measurement of the BLM capacitance. Hence the bifacial tension can be determined from a simultaneous measurement of P and the membrane capacitance (proportional to the BLM area, see Fig. 29). It should be pointed out that in the latter method the presence of electrical charge across the BLM may have a profound effect on the bifacial tension of the membrane. Therefore, in case there is any doubt, the use of the first method is recommended.

The results of bifacial tension measurements on several BLM systems are given in Table IX.

C. The Thermodynamic Quantities of BLM

With the experimental arrangement described above, the bifacial tension of a BLM as a function of temperature may be determined. From the γ-T data, various thermodynamic quantities can be calculated [70]. The basic equations are given below. It should be mentioned that the equations derived for monolayers are assumed to be valid for the BLM systems.

The interfacial pressure (π_i) for the BLM system is given by

$$\gamma_{BLM} - \gamma_{O/W} = \Delta F_i = -\pi_i \tag{93}$$

where γ_{BLM} is the bifacial tension, and $\gamma_{O/W}$ is the interfacial tension between pure solvent and water. Equation (93) also represents the decrease in the bifacial energy as a result of the BLM formation. The entropy of BLM formation can be written as

$$\frac{d(\Delta F_i)}{dT} = -\frac{d\pi_i}{dT} = -\frac{d\gamma_{BLM}}{dT} = S \tag{94}$$

and the heat of BLM formation is given by

$$H = \frac{\delta(\Delta F_i)/T}{\delta(1/T)} \tag{95}$$

The other important thermodynamic equation is the Gibbs adsorption isotherm, which may be written as

$$\Gamma = -(1/RT) \ (d\gamma_{BLM}/d \ln a) \tag{96}$$

where Γ is the interfacial excess, R and T are the gas constant and the absolute temperature, respectively, and a is the activity of the solute. The usefulness of these equations as applied to the BLM systems has been demonstrated [47].

VII. ELECTRICAL EXCITABILITY IN BLM

A. General Considerations

A potential difference in the order of millivolts has been found in all living animal and plant cell membranes. In the case of the nerve cell membrane, a comparatively large potential in the range of 50 to 90 mV is usually found, with the inside of the membrane negative with respect to the outside. Upon stimulation, e.g., by applying an electrical pulse to nerve tissue, a variation of membrane potential as a function of time is generally observed

TABLE IX

Interfacial Tension of Bimolecular Lipid Membranes at 25° C

BLM formed from:	Aqueous solution	γ_i, dyn/cm
Cholesterol–dodecane	0.1 N NaCl + HDTAB[a]	0.15 ± 0.05
Cholesterol–dodecane	H_2O + HDTAB[a]	0.65 ± 0.08
Egg lecithin–dodecane	0.1 N NaCl	0.90 ± 0.10
Cholesterol–DAP[a] -dodecane	0.1 N NaCl	1.1 ± 0.1
Oxidized cholesterol–octane	0.1 N NaCl	1.9 ± 0.5
Cholesterol–DODP[a]-dodecane	0.1 N NaCl	3.9 ± 0.5
Cholesterol–DODP[a]-dodecane	H_2O	5.7 ± 0.8
Chloroplast extract	0.1 N NaCl	3.8 ± 0.2

[a] HDTAB = hexadecyltrimethylammonium bromide; DAP = dodecyl acid phosphate; DODP = dioctadecyl phosphite.

under appropriate conditions. This membrane potential variation, known as the "action potential," may be detected by means of micropipet electrodes. The action potential is triggered by a small depolarizing current which causes local depolarization across the membrane. A highly readable account of nerve excitation has been given by Hodgkin and Huxley [71].

Action potentials and their associated phenomena can also be demonstrated in experiments with suitably modified BLM. Before presenting the technical details in generating electrical excitability in BLM, a brief account of electrophysiological background and current findings is in order.

The first coherent theory of electrical transients of excitable tissues was proposed by Bernstein almost 70 years ago. According to Bernstein's theory, the membrane of living cells has a low permeability to ions but is selective to K^+ ions in its normal (resting) state. During the passage of a nerve impulse, however, the membrane momentarily loses its selective property, and the resting potential drops to zero owing to about equal concentrations of ions (Na^+ + K^+) across the membrane. The movement of K^+ ions (and Na^+ ions) would initiate local depolarization of the adjacent region of the membrane. Thus the action potential is made to travel by a self-repetitive process along the length of the tissue. Although today the Bernstein membrane theory has been shown to be inadequate, its basic postulates are still valid and have been modified and extended by Hodgkin and Huxley [71]. The now classical work of Hodgkin and Huxley demonstrated that the action potential can be quantitatively related to the membrane potential and trans-membrane conductance due to the electrical currents carried by K^+ and Na^+ ions. Central to the Hodgkin and Huxley theory is the initial entry of Na^+ ions upon stimulation. It recently has been found that, for the squid giant axon at least, the electrical excitability does not depend upon the concentrations of K^+ and Na^+ ions. Apparently the electrical excitability of squid giant axon requires only an appropriate electrolyte gradient (e.g., Li^+, Rb^+, or Cs^+ together with a divalent ion such as Ca^{2+} can be used externally). Further, the membrane is still excitable even when the internal K^+ ions are replaced with Na^+. This latter finding is at variance with the Hodgkin-Huxley theory but has been explained by Tasaki et al. in terms of an ion exchange reaction involving membrane conformation [72]. These new findings imply that our understanding concerning membrane excitability is still incomplete and that squid giant axon might still be too complex a system to be amenable to a detailed analysis. Perhaps a simpler system such as a BLM may be a useful tool for further understanding.

B. Action Potentials Induced in BLM

The method given in this section is essentially that developed by Mueller et al. [73]. They had observed that a proteinaceous material (from an old bottle of egg white manufactured by Difco) dramatically lowered the BLM resistance, and in addition induced electrical excitability to dc electrical stimulation [73]. The compound(s) known collectively as excitability-inducing material (or EIM) which was responsible for the interesting observation is still to be identified but is obtainable from Aerobacter cloacae fermented egg white. A method for its partial purification has been given [74]. For the interested readers who wish to make their own EIM, a simplified procedure is given in Appendix E.

For the purpose of observing electrical transient phenomena in BLM, the membrane may be formed from brain lipid extract (see Appendix A), oxidized cholesterol (see Appendix B), or oxidized cholesterol + DAP mixture [32]. Since success in observing these phenomena requires special attention to details, the following procedure is recommended:

(1) A BLM is formed in distilled water buffered at pH 6.8 with 5 X 10^{-3} M histidine (22-26° C). The BLM resistance should be greater than $10^8 \, \Omega\text{-cm}^2$.

(2) Add 0.1 ml of 5% active egg white prepared according to the procedure given in Appendix E. The chamber to which EIM (active egg white) is added should be stirred with a magnetic stirrer and is designated as the "inside." To prevent bulging of the membrane, an equal amount (0.1 ml)

is withdrawn from the inside. This operation is best accomplished with an
Oxford Sampler (Oxford Laboratories, San Mateo, Calif.).

(3) The resistance of the BLM is monitored every minute until it decreases to about 1 to 3 X 10^6 Ω-cm^2 or less. The time required for this to
happen depends upon the efficiency of stirring and the EIM activity, but it
usually takes about 5 min. If the membrane resistance does not decrease
sufficiently within 10 min, another 0.1 ml of EIM solution may be added.

(4) Add 0.01 ml of 4 M KCl per ml to the "inside" solution. A further
reduction in membrane resistance may be observed. In addition, a transmembrane potential of 40 to 60 mV will develop with the inside negative.
This transmembrane potential is dependent upon the ratio of K^+ ion across
the membrane.

(5) Apply rectangular current pulses of 5-sec duration. This is done by
passing current through a series resistance at least 10 times that of the membrane. The polarity of the applied current is set to decrease the transmembrane potential.

(6) Continue applying current pulses at 10-sec intervals with voltages of
increasing current strength until an electrical transient is produced. This
electrical transient phenomenon may be observed repeatedly after allowing a
10-sec lapse (Fig. 31).

C. Suggestions for Plotting the Data

The current/voltage (I/V) curve for an EIM-treated BLM is usually
nonlinear and exhibits interesting negative resistance characteristics. The
data for a typical I/V plot may be obtained by measuring steady-state transmembrane potential (V_m) with known applied voltage (V_a) and series resistance (R_s). The membrane current (I_m) is calculated from the relation
$I_m = (V_a - V_m)/R_s$. Since R_s is usually 10 times larger than membrane
resistance, the I/V curve obtained in this manner is essentially under constant
current conditions.

This interesting I/V curve (see Fig. 32) can be explained by a simple
model in terms of the theory of reaction rates [75]. The suggested model
assumes that: (1) A number of pores (or channels) are formed by EIM;
(2) each channel has two gates, R_1 and R_2, connected in series; (3) the gates
can have either open or closed configurations; (4) the transition from open
configuration (denoted by a) to closed configuration (denoted by b) is dependent upon the energy difference, ΔE, between the two states and the transmembrane potential; (5) the transition between the two configurations follows
first-order kinetics; and (6) the effect of the membrane capacitance on
transition rates is negligible. With these assumptions in mind, a schematic
diagram showing an excitable membrane together with the stimulating circuit
is given in Fig. 32 [75]. The equation useful in constructing the theoretical
I/V curve is derived as follows:

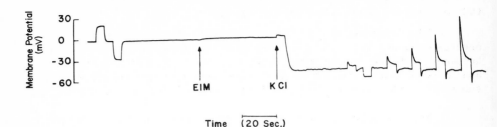

Time (20 Sec.)

Fig. 31. The development of conductance, resting potential, and delayed rectification in a bimolecular membrane. The membrane, 1.0 mm^2 in area, was formed at 25 °C separating two aqueous compartments containing 5 mM/liter histidine chloride buffer at pH 6.8. Test pulses were applied through series resistors which were usually kept higher than the membrane resistance so that the membrane current was constant during the pulse. The first test of the membrane resistance was made by applying ±80 mV through a series resistance of 2 X 10^8 Ω, which gave the two voltage pulses on the left. At the arrows, 0.1 cm^3 of an 0.05-gm/cm^3 solution of crude EIM and 0.1 cm^3 of a 2-M/liter KCl solution were added to one compartment of 4 cm^3 and stirred. Shortly thereafter, the membrane resistance began to fall and a resting potential appeared which was 50 mV negative on the side of the KCl, defined as the inside. Within 2 min there appeared strong delayed rectification, i.e., the membrane resistance and potential decreased during the applied constant outward currents. Under these conditions, the steady-state current/voltage (I/V) curve shows that the ratio of the membrane resistance (R = V/I) at the resting potential to that at zero potential can be as large as 1000 : 1 [76].

For gate R_1 (or R_2), let p be the fraction of gates in the high resistance state (b), and (1 - p) be the fraction in the low resistance state (a); then the rate of change of p is given by

$$dp/dt = k_1 (1 - p) - k_2 p \tag{97}$$

where k_1 is the rate constant for the forward transition and k_2 is the rate constant for the backward transition. The rate constants according to the theory of reaction rates are given by the following equations:

$$k_1 = A_1 \exp [- (E_1 + 11.5 V_m)/RT] \tag{98}$$

$$k_2 = A_2 \exp [- (E_2 + 11.5 V_m)/RT] \tag{99}$$

where A's are constants, 11.5 is the conversion factor, R is the gas constant, T is the absolute temperature, and V_m is given by

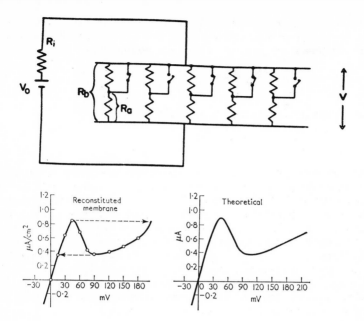

Fig. 32.   (A) Network representing an excitable membrane and stimulating circuit.   Each penetrating absorbate molecule is assumed to form a channel which can exist in two states having either a low (R_a) or high (R_b) resistance.   A stimulating voltage (V_O) is supplied from a battery through an input resistor R_i.   V is the recorded transmembrane potential.   (B) Current/voltage curves calculated from equations given in the text using selected constants to fit the data of the reconstituted membrane [75].

$$V_m = \frac{V_a}{1 + (R_s N/R_a)\{1 + p[(R_a - R_b)/R_b]\}} \qquad (100)$$

in which V_a is the applied voltage given earlier and N is the total number of channels.   Since $\Delta E = E_2 - E_1$, Eq. (97) can be integrated:

$$t = \int_{p_0}^{p} \frac{dp}{k_1(1-p) - k_2 p} \qquad (101)$$

where P_0 is the equilibrium value of p at $V_a = 0$ and is obtained from

$$P_0 = \frac{A_1 \exp(-E_1/RT)}{A_1 \exp(-E_1/RT) + A_2 \exp(-E_2/RT)} \qquad (102)$$

Equation (101) may be evaluated by numerous methods from plots of Eq. (97). From a knowledge of p as a function of t and V_a, the membrane resistance (R_m) at any time t and applied voltage can be calculated from

$$\frac{1}{R_m} = \frac{pN}{R_b} + \frac{N - pN}{R_a} \tag{103}$$

The membrane potential (V_m) at any time is given by

$$V_m = (R_m \cdot V_a)/(R_i + R_m) \tag{104}$$

The result of theoretical calculation of one case using assumed values for the constants of Eq. (102) is shown in Fig. 32 together with that obtained for an experimental BLM (treated with EIM). In recent experiments it has been found that repetitive and rhythmic firing may also be observed in suitably modified BLM [76]. For instance, a large cyclic peptide of known amino acid composition can replace EIM in generating all of the electrical transients mentioned above. Further experiments along these lines should be fruitful since at present there is little experimental evidence to suggest what the mechanisms might be in precise physicochemical terms. It seems probable, however, that the BLM-protein interaction is intimately connected with these voltage-controlled negative resistances.

VIII.   MEASUREMENT OF LIGHT-INDUCED CHANGES

The process of charge-carrier generation by light irradiation can be observed in BLM containing photoactive compounds such as chlorophyll and its derivatives. Experiments involving photoelectric effects in BLM are of interest, especially in relation to current theories of photosynthesis and visual excitation processes where highly organized lamellar structures are involved [77, 78].

The two most commonly studied photoelectric phenomena are the photovoltaic effect and photoconduction. Photovoltaic effect means that light produces a voltage across a barrier in the absence of an externally applied electric field. Photoconduction is the marked increase in electrical conductivity when a material is illuminated. Experimental details in observing these two interesting photoelectric effects in BLM are given in this section.

A. Photovoltaic Effects in BLM

The experimental arrangement used is the same as that shown in Fig. 33. A BLM is formed on the aperture in the usual manner. (A procedure of preparing a BLM-forming solution suitable for this experiment is given in Appendix C.) An illuminated slit is focused on the BLM, producing charge carriers in the BLM. The expected output voltage is of the order of several millivolts (the BLM system may be suitably modified, however, to produce a voltage greater than 50 mV). With the aid of neutral density filters or

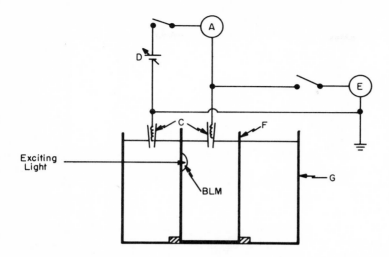

Fig. 33. Experimental setup for measuring the photoelectric effects of BLM. A-picoammeter, C-reversible electrodes, D-variable voltage source, E-electrometer, F-Teflon chamber with a small aperture, G-Pyrex glass chamber.

other light-modulating devices, the output photovoltages can be measured as a function of incident light intensity. Instead of using white light, the photo-voltaic effect of a BLM can also be investigated as a function of wavelength by use of a monochromator [84].

It is also possible to record variations of open circuit photovoltage in response to alternate illumination and darkness. A simple camera shutter or a rotating disk is used to interrupt the light beam. Figure 34 shows typical responses of a BLM to illumination cycles.

B. Photoconductivity in BLM

The apparatus used for measuring photoconductivity is illustrated in Figure 35. Both dark and photo-currents are measured directly with a picoammeter. From these measurements the open circuit photovoltage can be calculated from the equation:

$$V_{op} = -(kT/e) \ln[1 + (G_L/G_D)] \tag{105}$$

where V_{op} is the open circuit photovoltage, G_L and G_D are the photo- and dark currents, respectively, and k, T, and e have their usual significance.

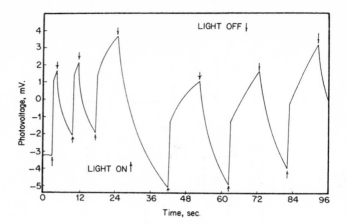

Fig. 34. Photovoltaic effect of a black lipid membrane produced from chloroplast pigments (spinach leaves) in n-octane. The curve illustrates membrane-voltage vs. time in response to photoactive light.

If simplifying assumptions are made, it can be shown that the charge carrier concentration, produced by illumination, obeys the equation

$$- dn/dt = (n - n_o)/\tau \tag{104}$$

where n is the light-generated carrier concentration, n_o is the equilibrium or dark concentration, and τ is the carrier lifetime. The carrier lifetime in the BLM may be determined using the well-known methods [85]. Instead of measuring the current directly as indicated in Fig. 36, the output voltage can also be measured. The circuit diagram generally used is illustrated in Fig. 36. The resistor used in series with the BLM must be 15-20 times larger than the BLM resistance in order to provide an effective constant current source. For this experiment an oscilloscope of sufficient sensitivity and a light pulse source must be used. The rate of decrease of photoconductivity with time in the BLM is recorded. The carrier lifetime is calculated from the following equation:

$$V_t = V_o \exp(-t/\tau) \tag{105}$$

where V_t and V_o are the recorded voltages at time t and zero, respectively. The carrier lifetime τ is the time required for the voltage in the pulse to drop to 1/e (=0.368) of its initial value. This quantity is easily obtained by plotting the output voltage as a function of time.

A typical photocurrent transient is shown in Fig. 37. It should be noted that the record was obtained with a thin lipid membrane [84]. In this particular case the photocurrent consists of two components. A semilogarithmic plot indicates that each component represents an apparent exponential decay, the lifetimes being <50 msec and 1.5 sec, respectively.

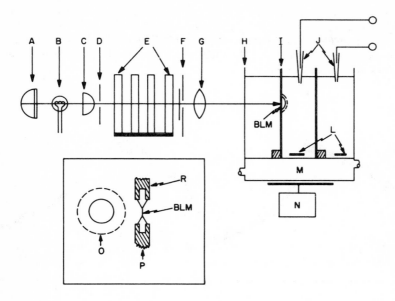

Fig. 35. Schematic diagram of experimental setup for measurement of photo-emf, photoconductivity, and dark conductivity of bilayer lipid membranes. A-concave mirror, B-500-W tungsten lamp, C-quartz collector, D-field diaphragm, E-filter carrier assembly, F-shutter, G-focusing lens, H-Pyrex glass chamber, I-Teflon chamber, J-electrodes, L-stirring bars, M-water jacket, N-magnetic stirrer, BLM-bilayer lipid membrane (or thin lipid membrane). Lower insert shows an enlarged view of the aperture in the Teflon chamber: O-front view, P-cross section (side view), R-Teflon support.

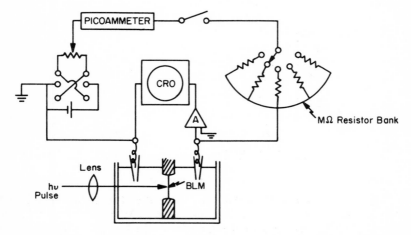

Fig. 36. Circuit diagram for the measurement of photoconductivity and the carrier lifetime in BLM (see text for details).

Using a set-up illustrated in Fig. 33, the time course of the light-induced photopotential in a chloroplast BLM is presented in Fig. 38. It should be mentioned that the photoresponse of pigmented BLM can be exceeding rapid (in the microsecond range). This and other light-initiated phenomena are being actively investigated.

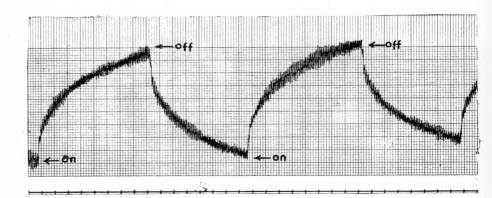

Fig. 37. Photocurrent of a thin lipid membrane periodically induced by blue light excitation (CuSO$_4$ solution filter used). The membrane contained chloroplast pigments. Aqueous solution: 0.1 M KCl + 10^{-4} M benzoquinone at pH 5.8 [84].

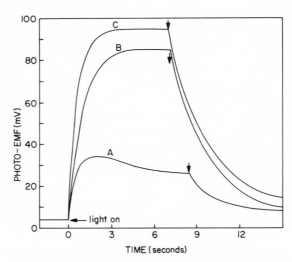

Fig. 38. Time course of the light-induced emf in a chloroplast BLM under various asymmetrical conditions. The BLM is formed in 0.1 M acetate buffer at pH 5.0. Exciting light is provided by a 150-watt tungsten lamp filtered through 8-cm CuSO$_4$ solution (5%). Downward arrows indicate light off. Curve A, 10^{-3}M FeCl$_3$ in the outer chamber (after 20 min); curve B, the same BLM with the inner chamber containing 1,4 dihydroquinone (after 1 hr); curve C, identical to A and B except that the pH of the inner solution has been adjusted to 8.4 with KOH.

APPENDIX A

Preparation of Brain Lipid BLM-Forming Solution

This is one of the successful solutions used by the original workers[34].
The method of preparation is as follows:

1. White matter from fresh beef brain obtained directly from the slaughter house is prepared by gross dissection.

2. Homogenize 25 g of white matter in 500 ml of $CHCl_3$-CH_3OH mixture (2:1, v/v) for 5 min at room temperature.

3. Clarify by centrifuging and by filtering. The resulting solution is about 1% in lipid content.

4. Add 20 ml of distilled water per 100 ml of lipid solution obtained in step 3 and shake to form an emulsion.

5. Lyophilize or evaporate to dryness under vacuum.

6. Dissolve the residue in 50 ml of $CHCl_3$ and filter through three layers of filter paper.

7. Add 1/2 volume of methanol to the solution from step 6.

8. Equilibrate solution from step 7 with 1/5 volume of 0.1 N NaCl by vigorous shaking.

9. Centrifuge and discard supernatant and interfacial fluff.

10. Filter lower phase three times (fresh filter paper each time).

11. Add 1/3 volume of methanol to filtrate. This gives a lipid solution of about 2:1 (v/v) in chloroform-methanol.

12. Store lipid solution obtained in step 11 in 1-ml lots in a freezer. This solution is stable for about 8 to 10 weeks.

13. The actual BLM-forming solution is prepared by adding 400 mg of α-tocopherol and 30 mg of cholesterol to 1 ml of solution from step 11 or 12. This BLM-forming solution, if kept in an ice bath or a refrigerator when not in use, may remain usable for about 5 days.

APPENDIX B

Preparation of Oxidized Cholesterol BLM-Forming Solution ·

General Principle

Freshly recrystallized cholesterol from commercial sources is oxidized in n-octane by bubbling molecular oxygen through the solution at its boiling point. The resulting colorless solution may be used directly for BLM formation.

Reagents

Cholesterol (Eastman, Rochester, N. Y.)

n-octane (practical grade, Eastman)

Ethanol (absolute)

Oxygen (compressed gas, Matheson)

Nitrogen (extra-dry compressed gas)

Equipment

Round-bottom flask, three-neck, 1000-ml

Heating mantle

Variable transformer (Variac)

Condenser, reflux

Gas-dispersing tube, fritted glass (medium porosity)

Flow meter (tube size R-2-15-AAA, Matheson)

Thermometer, 0-150°C

Funnel, fritted glass (medium porosity)

Beakers

Flask, filter

Heating plate with magnetic stirring provision

Stirring magnet, Teflon-coated

Procedure

1. Clean all glassware thoroughly (wash in hot detergent solution, rinse with distilled water, then acetone, and preferably dry in an oven at 120°C).

2. Bubble nitrogen through 300 ml of absolute ethanol for 30 min (via gas-dispersing tube).

3. Add 25 g of cholesterol to 250 ml of nitrogenated ethanol.

4. Warm the mixture to about 70° C until a clear solution results.

5. Cool to 5° C and let stand for 1 hr (in a refrigerator).

6. Filter the above solution with suction.

7. Repeat steps 2-6, except use 180 ml of nitrogenated ethanol in step 3.

8. Lyophilize the combined filtered masses overnight. The residue should be white and fluffy.

9.   Add 12 g of the material from step 8 to 300 ml of n-octane in a
     1-liter flask.

10. Heat the mixture to boiling ($\sim 125^\circ$ C) and bubble O_2 at a rate of
     100 to 125 cm^3/min for 5.5 to 6.0 hr.

11. Cool to room temperature. The solution should be colorless with a
     white precipitate.

12. Pipet off the supernatant for BLM formation.

APPENDIX C

Preparation of Chloroplast BLM-Forming Solution

A number of methods for extracting chloroplast lipids and pigments from
fresh spinach leaves have been tried. The procedure given in the flow chart
has been found to yield an extract suitable for BLM formation [83].

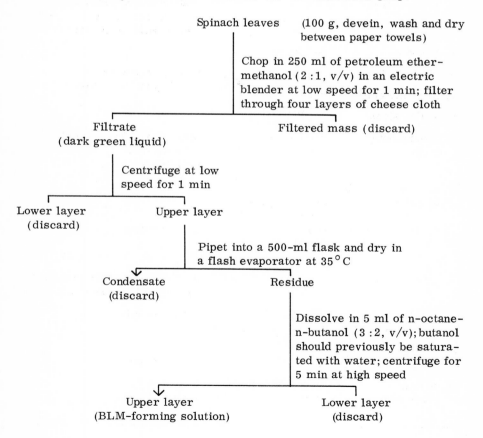

APPENDIX D

Parts List for an Inexpensive Setup

Item	Model and catalog no.	Manufacturer
1. Electrometer with multi-speed recorder	Heath EUW-301M	Heath Instrument Co. Benton Harbor, Mich.
2. pH text box	Heath EUA-20-12	Heath Instrument Co. Benton Harbor, Mich.
3. Electrodes (two)	Fiber junction calomel (39270)	Beckman Instrument Co. 25511 Southfield Road Southfield, Mich. 48075
4. Light source	Unitron LS	Unitron Instrument Co. 66 Needham Street Newton Highlands, Mass.
5. Eyepiece	No. 70,266	
6. Magnetic bars	No. 40,418	Edmund Scientific Co. Barrington, N.J. 08007
7. Magnetic cord	No. 40,875	
8. Teflon stock	No. 70,967	
9. Synchronous motor	No. H7-401 (60 RPM)	Harbach and Rademan, Inc. 1204 Arch St. Philadelphia, Pa.
10. a. Cell assembly	Design	
b. Glass cup	Design suggested in text	
c. Teflon beaker	10 ml	Will Scientific Co. Box 63 Ann Arbor, Mich. 48107
11. Coaxial cable		Amphenol-Borg Electronics Co. Broadview, Chicago, Ill.
12. Electrical connectors	No. 202HH Bakelite-insulated phone tip jack	Obtain locally
13. Sable hair brush	Artists'	Obtain locally
14. Syringe dispenser	Model PB-600-1 (100 μl)	Hamilton Co. Whittier, Calif.

APPENDIX E

Preparation of Excitability-Inducing Material (EIM)

Excitability-inducing material (EIM), although as yet unidentified, lowers the resistance of a BLM by several orders of magnitude.   EIM is of special interest in that it causes the BLM to exhibit a highly nonlinear current/voltage curve with negative resistance characteristics.   This very unusual material can be easily prepared with the following simplified procedure [24].

Equipment

Drying oven (preferably with forced air draft); shallow pan; electric mixer.

Reagents

Fresh eggs (2 dozen); thioglycollate medium (Difco Laboratories)

Aerobacter cloacae (ATCC #961, 962, or 10699 obtainable from American Type Culture Collection, Rockville, Md.)

Procedure

A.   Bacterial inoculum

1.   Prepare the thioglycollate medium according to the manufacturer's directions.

2.   Inoculate the medium with a culture of Aerobacter cloacae.

3.   Incubate the inoculated medium at 37°C overnight.

4.   The resulting growth may be stored at 5-10 °C for 2 weeks.

B.   Active egg white

1.   Separate egg white from yolk in a large beaker.

2.   Beat the egg whites at a low speed with the mixer until uniformly whipped (a few minutes).

3.   Adjust the pH of the egg whites to 7.4.

4.   Inoculate the egg whites with Aerobacter cloacae inoculum (from A).

5.   Ferment the mixture for about 30 hr at 37°C.

6.   Dry the fermented mixture in a shallow pan at 50°C overnight.

7.   Store the resulting flakes (containing EIM) in an airtight bottle for ready use.   (EIM prepared in this manner is known to be active for several years when at room temperature.)

8.   A 5% solution of active egg white in 0.1 M NaCl is freshly prepared when needed.   (Note: To induce electrical excitability only about 0.1 to 1 mg of active egg white is needed per membrane.)

REFERENCES

1.   J. A. Castleden, J. Pharm. Sci., 58, 149 (1969).

2.   L. Rothfield and A. Finkelstein, Ann. Rev. Biochem., 37, 480 (1968).

3.   D. A. Haydon, in Membrane Models and the Formation of Biological Membranes, North-Holland, Amsterdam, 1968, pp. 91-96.

4.   H. T. Tien and A. L. Diana, Chem. Phys. Lipids, 2, 55 (1968).

5.   E. Gorter and F. Grendel, J. Exptl. Med., 41, 439 (1925).

6.   E. Overton, Vjsehr. Naturf. Ges Zürich, 40, 149 (1895).

7.   H. Davson and J. F. Danielli, The Permeability of Natural Membranes, University Press, Cambridge, England, 1952.

8.   J. D. Robertson, Progr. Biophys., 10, 343 (1960).

9.   A. A. Benson, J. Am. Oil Chemists' Soc., 43, 265 (1966).

10. R. Beutner, J. Am. Chem. Soc., 36, 2040 (1914).

11. L. Michaeli and A. A. Weech, J. Gen. Physiol., 12, 55 (1928).

12. T. Teorell, ibid., 21, 107 (1937).

13. K. Sollner, Ann. N. Y. Acad. Sci., 57, 177 (1953), and references therein.

14. J. M. Tobias, D. P. Agin, and R. Pawlowski, J. Gen. Physiol., 45, 989 (1962).

15. A. D. Bangham, Progr. Biophys., 18, 29 (1968).

16. D. Papahadjopoulos and N. Miller, Biochim. Biophys. Acta, 135, 624 (1967).

17. R. B. Dean, Nature, 144, 32 (1939).

18. D. O. Shah and J. H. Schulman, J. Lipid Res., 8, 215 (1967).

19. J. W. McBain, Advances in Colloid Science, Vol. 1, Wiley (Interscience), New York, 1942.

20. J. T. Davies and E. K. Rideal, Interfacial Phenomena, Academic, New York, 1961.

21. H. B. Bull, An Introduction to Physical Biochemistry, F. A. Davis Co., Philadelphia, Pa., 2nd ed., 1971.

22. For a recent monograph, see G. L. Gaines, Insoluble Monolayers at Liquid-Gas Interfaces, Wiley (Interscience), New York, 1966.

23. P. Mueller, D. O. Rudin, H. T. Tien, and W. C. Wescott, Symposium on the Plasma Membrane, New York City, December, 1961; subsequently published in Circulation, 26, 1167 (1962).

24. P. Mueller, D. O. Rudin, H. T. Tien, and W. C. Wescott, in Recent Progress in Surface Science, Vol. 1, Academic, New York, 1964, Chap. 11.

25. H. T. Tien and E. A. Dawidowicz, J. Colloid Interface Sci., 22, 438 (1966).

26. J. F. Danielli, Colston Papers, 7, 1 (1954).

27. D. E. Green and J. Perdue, Proc. Natl. Acad. Sci., U. S., 55, 1295 (1966).

28. J. Folch and M. Lees, J. Biol. Chem., 191, 807 (1951).

29. R. E. Howard and R. M. Burton, J. Am. Oil Chemists' Soc., 45, 202 (1968).

30. H. T. Tien, S. Carbone, and E. A. Dawidowicz, Nature, 212, 718 (1966).

31. T. Hanai, D. A. Haydon, and J. Taylor, Kolloid Z., 195, 42 (1964); Proc. Roy. Soc. (London), A281, 377 (1964).

32. H. T. Tien, J. Phys. Chem., 71, 3395 (1967).

33. H. T. Tien and A. L. Diana, J. Colloid Interface Sci., 24, 287 (1967).

34. P. Mueller, D. O. Rudin, H. T. Tien, and W. C. Wescott, J. Phys. Chem., 67, 534 (1963).

35. C. Huang, L. Wheeldon, and T. E. Thompson, J. Mol. Biol., 8, 148 (1964).

36. R. E. Howard, Ph.D. Thesis, Washington University, 1968.

37. C. Huang and T. E. Thompson, J. Mol. Biol., 15, 539 (1966).

38. T. E. Andreoli, J. A. Bangham, and T. Tosteson, J. Gen. Physiol., 50, 1729 (1967).

39. A. Cass and A. Finkelstein, ibid., 50, 1765 (1967).

40. H. J. Vandenberg, J. Mol. Biol., 12, 290 (1965).

41. R. Pagano and T. E. Thompson, Biochim. Biophys. Acta, 144, 666 (1967).

42. L. M. Tsofina, E. A. Liberman, and A. V. Babakov, Nature, 212, 681 (1966).

43. B. V. Derjaquin, A. S. Titijevskaia, I. I. Abricassova, and A. D. Malikna, Discussions Faraday Soc., 18, 24 (1954).

44. A. Scheludko, Advan. Colloid Interface Sci., 1, 391 (1967).

45. M. Takagi, K. Azuma, and V. Kishimoto, Ann. Rept. Biol., 13, 107 (1967).

46. F. A. Henn and T. E. Thompson, J. Mol. Biol., 31, 227 (1968).

47.   H. T. Tien, J. Gen. Physiol., 52, 125s (1968).

48.   K. J. Mysels, K. Shinoda, and S. Frankel, Soap Films, Pergamon, New York, 1959.

49.   V. Luzzati and F. Husson, J. Cell. Biol., 12, 207 (1962).

50.   A. W. Adamson, Physical Chemistry of Surfaces, Wiley, New York, 1967.

51.   H. T. Tien, J. Theoret. Biol., 16, 97 (1967).

52.   O. S. Heavens, Optical Properties of Thin Solid Films, Butterworths, London, 1955.

53.   A. Vasicek, Optics of Thin Films, North-Holland, Amsterdam, 1960.

54.   J. M. Corkill, J. F. Goodman, D. R. Haisman, and S. P. Harrold, Trans. Faraday Soc., 57, 821 (1961).

55.   E. M. Duyvis, Thesis, University of Utrecht, 1962.

56.   E. A. Dawidowicz, Ph. D. Thesis, Northeastern University, 1968.

57.   R. J. Cherry and D. Chapman, J. Mol. Biol., 30, 551 (1967).

58.   F. S. Sjostrand, Radiation Res. Suppl., 2, 349 (1960).

59.   W. Stoeckenius, in Principles of Bimolecular Organization, Ciba Sym., Little, Brown, Boston, 1966, pp. 418-441.

60.   F. A. Henn, G. L. Decker, J. W. Greenawalt, and T. E. Thompson, J. Mol. Biol., 24, 51 (1967).

61.   T. Hanai, D. A. Haydon, and J. Taylor, J. Gen. Physiol., 48, 59 (1965).

62.   P. Läuger, W. Lesslauer, E. Marti, and J. Richter, Biochim. Biophys. Acta, 135, 20 (1967).

63.   E. J. W. Verwey and J. T. G. Overbeek, Theory of Stability of Lyophobic Colloids, Elsevier, Amsterdam, 1948.

64.   P. Lauger, J. Richter, and W. Lesslauer, Ber. Buns. Phys. Chem., 71, 906 (1967).

65.   G. Eisenman, D. O. Rudin, and J. Casby, Science, 126, 831 (1957).

66.   C. Lippe, J. Mol. Biol., 35, 635 (1968).

67.   R. C. Bean, W. C. Shepherd, and H. Chan, J. Gen. Physiol., 52, 495 (1968).

68.   C. Huang and T. E. Thompson, J. Mol. Biol., 15, 539 (1966).

69.   H. T. Tien and H. P. Ting, J. Colloid Interface Sci. , 27, 702 (1968).

70.   H. T. Tien, J. Phys. Chem. , 72, 2723 (1968).

71.   A. L. Hodgkin, The Conduction of Nervous Impulses, C. Thomas,
       Springfield, Ill. , 1964.

72.   I. Tasaki, Nerve Excitation, C. Thomas, Springfield, Ill. , 1968.

73.   P. Mueller, D. O. Rudin, H. T. Tien, and W. C. Wescott, Nature,
       194, 979 (1962).

74.   L. D. Kushnir, Biochim. Biophys. Acta, 150, 285 (1968).

75.   P. Mueller and D. O. Rudin, J. Theoret. Biol. , 4, 268 (1963).

76.   P. Mueller and D. O. Rudin, Nature, 213, 603 (1967).

77.   M. Calvin, Rev. Mod. Phys. , 31, 147 (1959).

78.   G. Wald, Science, 162, 238 (1968).

79.   H. T. Tien, J. Phys. Chem. , 72, 4512 (1968).

80.   A. A. Lev, V. A. Gotlib, and E. P. Buzhinsky, Zh. Evolyu. Biokhim.
       Fiziol. , 2, 109 (1966).

81.   P. Mueller and D. O. Rudin, Biochem. Biophys. Res. Commun. ,
       26, 398 (1967).

82.   A. A. Lev and E. P. Buzhinsky, Zh. Evolyu. Biokhim. Fiziol. , 9,
       102 (1967).

83.   H. T. Tien, W. A. Huemoeller, and H. P. Ting, Biochem. Biophys.
       Res. Commun. , 33, 207 (1968).

84.   H. T. Tien, in The Chemistry of Bio-Surfaces, Marcel Dekker, Inc.,
       New York, 1971, Chap. 6.

85.   S. M. Ryvkin, Photoelectric Effects in Semiconductors, Consultants
       Bureau, New York, 1964.

Chapter V

METHODS USED IN THE VISUALIZATION OF CONCENTRATION GRADIENTS

Carel J. van Oss

Department of Microbiology
School of Medicine
State University of New York at Buffalo

I. INTRODUCTION

In diffusion, sedimentation, and electrophoresis of dissolved substances, a concentration redistribution occurs between a transparent solution and its equally transparent solvent, which remains invisible to the naked eye. For a quantitative and nondisruptive study of these phenomena, it is nevertheless desirable to visualize and to record these concentration redistributions. For that purpose two different optical principles can be applied:

(a) absorption of light of such a wavelength that the solute is opaque, while the solvent is transparent to it;

(b) refraction of light going through those places in the cell where a concentration gradient (and thus a refractive index gradient) exists.

213

The light absorption method can yield concentration versus distance curves and is useful when it is necessary to work with very low solute concentrations (less than 0.1%) or when interactions of two or more different kinds of solute are to be studied, which have maximum absorptivities at different wavelengths.

Various light refraction methods can yield concentration vs. distance curves (the interference and the prismatic cell methods), change in concentration versus distance curves (the scale and moiré methods, as well as the schlieren method), and the second derivative curves (through an extension of the scale method).

II. THE CONCENTRATION FUNCTION AND ITS FIRST TWO DERIVATIVES

In any system in which a concentration redistribution is created, whether or not it is caused or influenced by an electric or a gravitational field, diffusion inevitably starts and inexorably follows its course. Thus, even when diffusion is not the prime object of study of a phenomenon involving a concentration redistribution, its occurrence, because it cannot be avoided, in all cases has to be taken into account. Therefore, unidimensional diffusion (with a constant diffusion coefficient) is treated here as the simplest, as well as the most fundamental, phenomenon illustrative of the concentration function.

The concentration function, representing diffusion in a rectangular cell, and its first two derivatives are shown in Fig. 1. The curves shown represent solutions [1] of Fick's second law [2]:

$$\frac{\delta C}{\delta t} = D \frac{\delta^2 C}{\delta x^2} \tag{1}$$

When information on the total amount or the concentration of solute is desired, the integral curve can be quite useful. In electrophoresis and in sedimentation studies, the first derivative curve is the most important, as the migration rates of the tops of the peaks directly yield the mobilities (or the sedimentation rates) of the various fractions. Finally, for diffusion studies the second derivative curve can be of great use, as, in accordance with Fick's second law [1], it also depicts the change of concentration with time vs. distance in the cell. Thus, the migration of the maximum and minimum of the second derivative curve (which represent the points of greatest change in concentration with time) can directly yield the diffusion coefficient of the diffusing solute.

III. REFRACTION OF LIGHT IN A CONCENTRATION GRADIENT

The refraction of light when it crosses a density or concentration gradient (a refraction which is, paradoxically, strongest when the light rays can cross the gradient perpendicularly) was first treated by Huygens in 1690 [3] and later by Wollaston in 1800 [3a]. Atmospheric refraction was much

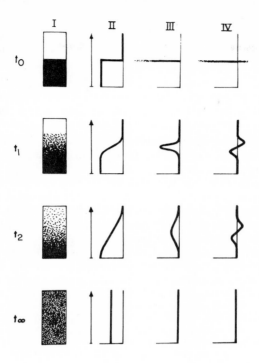

Fig. 1. The concentration function (representing diffusion in a rectangular cell) and its first two derivatives. The four rows represent, from top to bottom, the start of the diffusion (t_0), an early stage (t_1), a later stage (t_2), and the final equilibrium stage (t_∞). The cells in column I show a front view of the cell, illuminated from behind with light of a wavelength that is strongly absorbed by the solute. Column II shows the concentration of solute plotted against (vertical) distance in the cell. Column III depicts the curve of change in concentration versus distance. These curves depict, in the case of an ideal concentration–independent diffusion, a Gaussian error distribution. Column IV illustrates the second derivative, or the change in change in concentration versus distance.

studied around the turn of the 18th century, with a view to elucidate the phenomenon of the mirage, or Fata Morgana, by Monge (1799), Laplace (1805), Biot (1809), and also by Grunert (1847), Bravais (1856), and Kummer (1860) (see Wiener [4]).

Up to 1893 these authors used diffusion in liquids as a model for the study of this type of refraction. Wiener [4] was the first deliberately to use the phenomenon of light refraction in concentration gradients for the study of diffusion. He thereby became the first investigator to make a major contribution to our knowledge of light refraction in concentration gradients.

Wiener was the first to realize that the "error," or skewness, which was
demonstrated by Stefan [5] to be inherent in all refraction methods in gradients, could indeed form the base for a method to study the laws of that very
refraction.

The basic facts are illustrated in Fig. 2. This shows how light pencils,
arriving perpendicularly upon a plane-parallel walled glass vessel containing
a refractive index gradient, cross the vessel undeviated at the top, which
contains pure solvent, as well as at the bottom of the vessel, which contains
undiluted solution. In the large middle part, however, where there is a
concentration (and thus a refractive index) gradient, the light pencils are bent
in the direction of the higher refractive index. It can be easily shown that the
degree of this refraction (or light-bending) is proportional to the degree of
change in refractive index [4]. It is this general phenomenon upon which all
refraction methods of measuring concentration gradients are based. And all
methods to be discussed hereafter (except the Gouy method) have devices to
deflect the downward refraction sideways (in order to record it in a direction
separate from and perpendicular to x). The prismatic cell, the rayleigh
interference, the moiré, and the schlieren methods do it directly; the Jamin,
absorption, and scale methods, which involve a scanning or a plotting step,
do it indirectly.

It should be noted that for most solutes, and in particular for proteins
(on which much work has been published), the refractive index increment
Δn is remarkably proportional to the change in concentration Δc [6, 7], and,
at least for one protein (bovine serum albumin), this proportionality is known

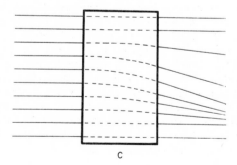

C

Fig. 2. Refraction of light by a refractive index gradient. Horizontal
light rays enter the cell (C) containing a concentration gradient from the
left. The rays on top and bottom cross the liquid with a uniformly low or
high refractive index, undeviated. But in the middle of the cell, where there
is a concentration (and thus a refractive index) gradient, the rays are bent
in the direction of increasing refractive index. This bending is proportional
to the change in refractive index, or, in other words, it is strongest where
the concentration gradient is sharpest.

to persist up to concentrations as high as 45% [8]. We shall therefore use Δc and Δn interchangeably.

IV. VISUALIZATION OF THE INTEGRAL FUNCTION

A. Prismatic Cell Methods

These methods are based upon the stronger refraction of light by a solution than by its solvent. When a gradient is established in a glass prism with vertical edges, the light rays going through the bottom of the prism (where the more concentrated solution is located) will be more deflected toward the base of the prism than light rays that go through the top of the prism, where there is only solvent. Thus, from top to bottom of the prism, the light rays will be more and more deflected and the image of a vertical light slit projected through the cell will form a concentration versus distance curve (see Fig. 3). This method was first described by Wild and Simmler in 1857 and modified by Johannisjanz in 1877 (see Wiener [4]). Although the method depicts the total concentration difference accurately enough, the intermediate part of the curve suffers from what Wiener [4] describes as the Stefan [5] skewness. This skewness is caused by the downward deflection of light rays caused by the refractive index gradient (see Sec. III and Fig. 2). This skewness is of course the more pronounced, the thicker the cell that contains the gradient becomes. Thus, through a prism, this downward skewness becomes worse as the vertical light slit is displaced toward the thickest side of the prism, and in all cases it causes a stronger downward deflection in the lower

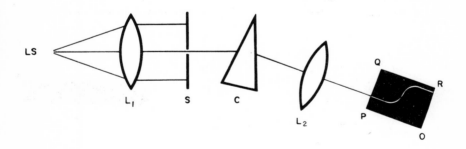

Fig. 3. The prismatic cell method. Top view, except for the projected image on the right, which is drawn as an elevation and should be imagined rotated 90° about the line PQ, so that it becomes perpendicular to the plane of the paper. LS–light source, L_1–collimating lens, S–vertical slit, C–prismatic cell with gradient, L_2–projecting lens, OPQR–screen or photographic plate.

part of the curve than in the higher part, which compromises its use as an exact depiction of the c, x curve. Nevertheless, for differential refractometry the method has its uses because it is quite simple, and it has at various times been applied and described by Thovert [10], Zuber [11], Debye [12], Kegeles [13], Labhart and Staub [14], and van Oss [15]. In ultracentrifugation in particular, when the concentration at the "plateau region" has to be determined, the total line deflection obtained with the prism cell method can give the desired information.

B.   Interference Methods

All three interference methods (Rayleigh, Jamin, and Gouy) to be discussed here yield the c, x curve in a more or less accessible form. All three of them make use of the fact that the "interference" bands, formed when a coherent, monochromatic light pencil is split and reunited [16], are displaced if the optical length of one of the pencils is changed, for instance, by making it cross a medium of different refractive index.

1.   The Rayleigh Interference Method

The Rayleigh interference method, first published in 1896 [17] and half a century later further developed by various authors [18-23], uses two vertical light ribbons from the same monochromatic source of wavelength λ, one of which crosses a cell containing only solvent, while the other crosses a cell which contains a concentration gradient of a given solute in that solvent (see Fig. 4). After each ribbon has passed a narrow vertical slit, they are reunited by a cylindrical lens into a system of narrow vertical interference fringes. The lower portion of these fringes is displaced sideways to an extent that is proportional to the increase in refractive index in the various layers of solution that were crossed. Thus, every fringe describes a c, x

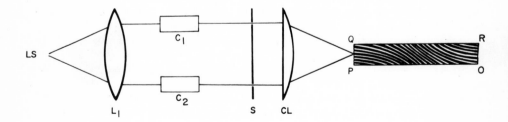

Fig. 4. The Rayleigh interference method. Top view, except for the projected image on the right, which is drawn as an elevation and should be imagined rotated 90° about the line PQ, so that it becomes perpendicular to the plane of the paper. LS-monochromatic light source, L_1-collimating lens, C_1-cell with solvent, C_2-cell with solute gradient (or vice versa), S-double vertical slit, CL-cylindrical lens with a vertical cylinder axis, OPQR-screen or photographic plate.

curve, but, a small difference in refractive index Δn in a cell of a thickness, a, gives rises to a large displacement j of the fringes, as, according to

$$j = a \, \Delta n / \lambda \tag{2}$$

j would amount to a shift of about 20 fringes in a 1-cm-thick cell for a value of Δn of 0.001, which corresponds to a difference in concentration of not more than 0.6% for most solutes. Since in most systems only eight fringes are visible, the c, x curve can only be followed on one and the same fringe if the gradient is a very faint one. In all other cases the total concentration difference can only be obtained by counting the fringes in the x direction and generally by including a fractional fringe [9]. Thus, the Rayleigh interference method, though it allows a very high resolution, tends to be fairly laborious, which is one of the major reasons that it is relatively rarely used even by the many processors of Spinco-E analytical ultracentrifuges that are equipped with these optics.

It nevertheless is easy to remedy this difficulty. Billick and Bowen [23] have significantly widened the scope of Raleigh interferometry by simply applying the principle, already discovered by Svensson [20d], that a perfectly bright, well-defined, and quite wide pattern of fringes can be obtained by using a system of multiple slits of specified dimensions, made with the help of a fine grating, which allows the utilization of the entire light source. In that manner most existing Rayleigh systems can be quite simply transformed so that the same fringe can be followed throughout the entire cell.

2. The Jamin Interference Method

The Jamin interference method, first published in 1856 [24], utilizes two identical plane-parallel glass plates placed some distance apart and exactly parallel to each other (see Fig. 5). The two faces that are oriented toward one another are half-reflecting; the two outer faces are totally reflecting. It will be clear from the diagram (Fig. 5) that the monochromatic light pencils, split by the first Jamin plate J_1 and reunited by plate J_2, would go through exactly the same optical pathlength if nothing were interposed between the two Jamin plates. As an image of the cell is projected on the screen unhindered, no more interference bands are seen than just one dark area diffusing into a white one. But when the two light pencils cross two media of different refractive indices, they will upon recombination give rise to a series of horizontal interference bands, distributed in such a way that the concentration difference between two bands is always the same. Thus, the close proximity of a number of fringes is always a sign of a sharp increase in concentration in that place, while widely spaced fringes are indicative of a very faint gradient. With these principles in mind, it is possible to train oneself to visualize the entire course of a concentration gradient by the simple inspection of its Jamin interference pattern. The exact c, x curve can be generated from the Jamin interference pattern in a simple manner, by plotting consecutive fringes on the ordinate against equal intervals on the abscissa (see Fig. 6). The use of graph paper facilitates this operation considerably.

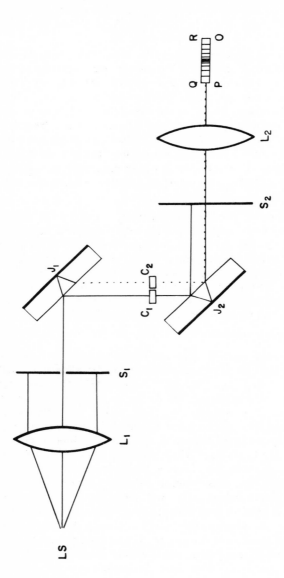

Fig. 5. The Jamin interference method. Top view, except for the projected image on the right, which is drawn as an elevation and should be imagined rotated 90° about the line PQ, so that it becomes perpendicular to the plane of the paper. LS-monochromatic light source, L_1-collimating lens, S_1-vertical slit, J_1-first beam-splitting Jamin plate (a plane-parallel glass plate with a half reflecting front and a totally reflecting rear surface), C_1-cell with solute gradient, C_2-cell with solvent, J_2-second beam-reuniting Jamin plate, S_2-vertical slit, mainly serving as screen for blocking off the unwanted beam, L_2-projecting lens, OPQR-screen or photographic plate.

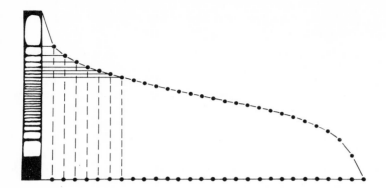

Fig. 6. Generation of c vs. x curve from Jamin fringes. As the concentration difference from one fringe to another is constant, the c vs. x
curve is obtained by plotting consecutive fringes on the ordinate against appropriate unit intervals on the abscissa. The fringes used here represent
the diffusion of 10% against 8% cadmium sulfate in water, photographed 120
min after starting the boundary. (Pattern and data obtained from Prospectus
505 on the LK 30 microelectrophoresis apparatus, Kern & Co. , Aarau,
Switzerland, which is, unfortunately, no longer being manufactured.)

        Labhart and Staub [25] were the first to use Jamin interference optics
in moving boundary electrophoresis. The great sensitivity of this optical
system (which is about the same as that of the Rayleigh system) makes it
possible to use microelectrophoresis cells (containing about 0. 5 ml of solution, with 5 mg of solute). This has the advantage of yielding complete elec-
trophoretic resolution of normal human serum in about 30 min. An additional
advantage of the use of such a small cell is that no elaborate cooling systems
are necessary. Lotmar [26] devised a variant on this microelectrophoretic
method, the apparatus for which was commercially available (Kern & Co,
Aarau, Switzerland) for a number of years, although its manufacture has
now been discontinued. The reason why Jamin interference optics never became more universally used, notwithstanding their undoubted advantage of
great sensitivity, was without much doubt the fact that they only yielded c, x
curves, and even those not directly but only after a certain amount of work.
And for the visualization of electrophoresis results with fairly complicated
protein mixtures like whole sera, the c, x curves somehow are not very informative to anyone without exceptional experience in this field. Although
the c, x curves permit the quantitation of every serum protein fraction, in
practice most people much prefer to view the $\partial c/\partial x$, x curve and to judge the
quantity represented by every fraction by estimating the surfaces under the
different peaks. The $\partial c/\partial x$, x curve can of course also be obtained from the
Jamin interference pattern (by plotting the reciprocals of the distances between consecutive fringes on the ordinate against their positions on the
abscissa), but that operation is much more laborious than plotting the c, x

curve, as it involves the extra steps of multiple measurements and calculations of the reciprocals.

The Jamin refractometer can also be used [27, 28] to obtain the c, x curve directly, without plotting. The Jamin plates must then each consist of two glass plates which can be slightly oriented away from the parallel, allowing vertical fringes to be formed by focusing, in much the same way as is done by the cylindrical lens in the Rayleigh system. The fringes thus obtained are more widely spaced than Rayleigh fringes, but they are not nearly as sharp. The actual optical system is quite complicated (for use in the ultracentrifuge it must, in addition, be provided with a light-chopping arrangement), and the system has not found widespread use.

3. The Gouy Interference Method

The Gouy interference method, based on qualitative observations made by Gouy [29] in 1880, was further developed with a view to quantitative diffusion measurements in the 1940's by Longsworth [30] and Kegeles and Gosting [31] and simultaneously but independently by Coulson et al. [32]. The method was further refined by Gosting and co-workers [33, 34] and by Ogston [35].

The method is based on the fact that when light rays cross a refractive index gradient, there are always on either side of the place of maximum refractive index change two light pencils that are parallel. All such pairs of parallel pencils can be focused onto one plane and, according to the differences in the light paths they have followed, they will then form a series of alternating dark and light horizontal fringes in that plane (see Fig. 7).

There is some resemblance between such a Gouy interference pattern and the lower half of a Jamin pattern (see Figs. 5 and 6), but unlike the Jamin interference pattern, the form of the Gouy pattern is not simply related to the form of the Gaussian distribution of concentration in the boundary [32]. For the simple case of purely Gaussian, single-component, concentration-independent diffusion, the mathematics have been worked out [32-35] (the method is probably only applicable to such a simple case anyway). Ogston et al. [32, 35] have applied the method for diffusion measurements at extremely short times after the formation of the boundary; Longsworth [32] and Gosting and co-workers [33, 34] utilized the method for measurements of diffusion after longer times. As Longsworth [32] already mentioned (see also Svensson [9]), the optical system necessary for the Gouy method can, with monochromatic light, make use of a schlieren optical system (see below), with the camera focused on a plane where the diagonal bar or phase plate is normally located (but in the absence of that bar and the cylindrical lens).

Ogston and co-workers [32] used an automatically moving photographic plate which allowed the recording of the entire diffusion process. Griffith and McEwen [36] adapted that method to an apparatus with existing schlieren optics (the Beckman-Spinco Model H electrophoresis and diffusion apparatus; not manufactured since 1968).

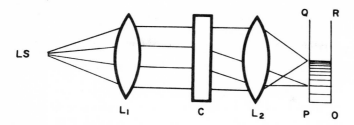

Fig. 7. The Gouy interference method. Elevation, except for the projected image on the right, which is a front view and should be imagined rotated 90° about the line PQ, so that it becomes perpendicular to the plane of the paper. LS-horizontal monochromatic light source, L_1 - collimating lens, C-cell with gradient, L_2 - projecting lens, OPQR-screen or photographic plate.

C. Absorption

The light absorption method for recording changes in solute concentration versus distance (c, x) was first used in analytical ultracentrifugation by Svedberg and his collaborators [37, 38]. The method is illustrated by column I of Fig. 1, in which a front view is shown of the cell, illuminated from behind with light of a wavelength that is strongly absorbed by the solute. From top to bottom the cell is shown, successively, at the start of the diffusion (t_0), at an early stage (t_1), at a later stage (t_2), and at the final equilibrium stage ($_\infty$). When photographic images of the cell, as depicted in column I of Fig. 1, are scanned in a densitometer, plots of optical density vs. distance in the cell are obtained, as illustrated in column II of the same figure. In all cases where Beer's law (asserting the proportionality of solute concentration and optical density) holds true, such plots represent c, x curves.

Like the Jamin interference method (see above), the absorption method allows the generation of c, x curves only after a certain amount of work; direct (or rather, almost direct) visualization is only possible when the optics are reinforced with an automatic scanning arrangement.

The earliest utilizations of the method all involved the indirect procedure of scanning photographs taken of the cell at various times at the appropriate wavelength. Many of the problems of photographic exposure, photometry, light filters, and development of the photographs were treated by Svedberg and Pederson [38] and later on by Schumaker and Schachman [39]. A fair number of the more recent Beckman-Spinco Model E analytical ultracentrifuges are equipped with a high intensity ultraviolet light source and monochromator, which can be used with an ultraviolet camera, as well as with a photoelectric scanning system. Schachman [40, 41] has been especially instrumental in developing an automatic split-beam photoelectric scanning system of the absorption image. This method can be used during the ultracentrifuge run; it circumvents the necessity of (and some of the difficulties inherent in) the analysis of photographic plates, and it allows the quasi-instantaneous production of c, x curves during the run.

Absorption optics have two intrinsic advantages over the other optical methods described in this chapter:

(1) Their very high sensitivity: Molecular weight measurements have been done with quantities of protein as low as 2×10^{-6} g [41].

(2) Their selectivity: While with refractive methods proteins and nucleic acids "look alike" [40], with absorption optics, using different wavelengths, such different components can be distinguished from one another in mixtures.

Nevertheless, absorption optics have not attained as wide a utilization as schlieren optics (see below), principally because they only readily allow the generation of c, x curves. (Although a first derivative generating attachment has been described [41], such devices are by no means in general use.)

Finally, some devices should be mentioned which describe c, x curves, after a fashion, by scanning electrophoresis or chromatography columns with monochromatic (generally ultraviolet) light and recording the light absorption (on logarithmic paper, to give a semblance of optical density) versus distance in the column. In this apparatus it is not actually the scanner that travels along the gradient; rather, the whole liquid column is pumped past the scanner. Apart from the convective disturbance that such a procedure tends to create, the major disadvantage of these contrivances is that the optical density (and thus the concentration) is depicted on logarithmic paper, so that no true c, x curve is directly obtainable. Instruments such as these are made by Instrumentation Specialties Co., Lincoln, Nebr. (Scanning Density Gradient Electrophoresis) and LKB-Produkter, Bromma, Sweden (Uvicord). All commercially available amino acid analyzers use the same principle for the analysis of the various fractions that come off a chromatography column (and indeed it is nowadays part of most automated analysis systems). The manner of recording in most of these analyzers suffers from the same drawback.

V. VISUALIZATION OF THE FIRST DERIVATIVE FUNCTION

A. Scale and Moiré Methods

1. Scale Method

This method is based on the demonstration of a refractive index gradient in an otherwise perfectly limpid medium by the distortion of a regular pattern when viewed through the gradient. The method was first applied by Wollaston in 1800 [3a] and later by Foucault [42], who tested telescopic mirrors for local irregularities by noting the distortions of the image of a net. Stefan[5] observed the lines on a piece of ruled paper through a salt gradient in water and noted that the lines apparently were displaced upward. This displacement was most pronounced when viewed through the location of the original solvent-solution boundary. Above this location the lines appeared thinner and closer together, below it thicker and wider apart. (See Fig. 10 for a photograph of such a distorted ruling and Fig. 8 for a graphical explanation of the phenomenon.)

We have already seen (Sec. III and Fig. 2) how light pencils are bent in the direction of the highest refractive index when crossing a refractive index gradient, and we have noted that the degree of this refraction is proportional to the degree of change in refractive index [4]. It follows that the degree of apparent upward displacement of equally spaced lines (running parallel to the plane of the boundary), when viewed through a refractive index gradient, is also proportional to the degree of change in refractive index. Then, the apparent displacement of the lines when viewed through the gradient must also be proportional to the change in concentration (see Sec. III), and, by just plotting the displacement of each line against its apparent location, the $\partial c/\partial x, x$ curve is produced.

This method was further elaborated by Lamm [43-47] (see also Refs. [48,49] and Svensson and Thompson's rather complete treatise [9]). In the 1920's and 1930's most of the experimental work on diffusion [50,51] and on analytical ultracentrifugation [38,52] was based on a fairly laborious plotting of $\partial c/\partial x, x$ curves by comparing distorted with undistorted "Lamm scales"

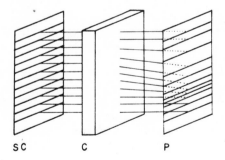

SC   C   P

Fig. 8. The scale method. SC-horizontally lined scale, C-cell with gradient, P-screen or photographic plate showing distorted scale. This is a simplified drawing: In reality the camera (or the naked eye) sees the distorted scale projected at a level behind the cell, not far from the original scale; the distorted scale image is then upside down as compared to the one shown here. When the amount of shift of every line is plotted against its original position, the $\partial c/\partial x, x$ curve is obtained. The same effect can be obtained in the moiré fringes formed when a photograph of the distorted scale (see Fig. 10) is laid on top of a photograph of the undistorted scale (see Fig. 9), without any laborious plotting (see Fig. 11).

(see Meyerhoff [53] and Schumacher [54] for attempts to automatize this operation). Notwithstanding the high accuracy of the scale method, it is hardly surprising that less tedious methods were actively sought and were eventually found in improved schlieren methods (see Sec. V.B).

2. Moiré Method

Nevertheless, the tedium of plotting line displacements for obtaining $\partial c/\partial x, x$ curves with the scale method can be quite easily avoided. By simply placing the photographic slides of the distorted and the undistorted scale at an appropriate angle on top of one another, the resulting moiré interference fringes form multiple $\partial c/\partial x, x$ curves [15, 55, 56] (see Figs. 9-11). Or, if one wishes, by placing a scale behind the gradient and another scale at a slight angle in front of it, multiple $\partial c/\partial x, x$ curves can be viewed directly (Fig. 2 in Ref. [15]).

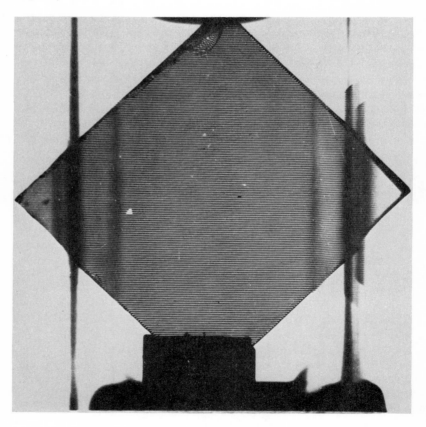

Fig. 9. Photograph of a grating with 150 lines/in. The black lines have the same thickness as their intervals. The photograph was taken through a cell filled with solvent (water).

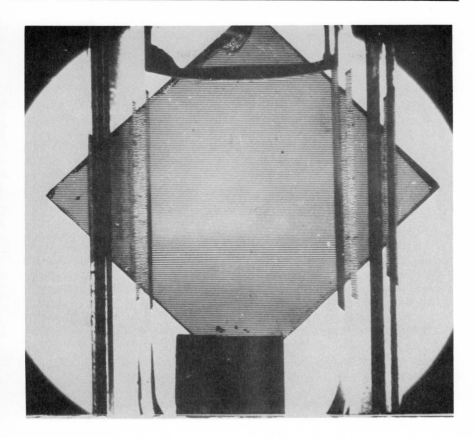

Fig. 10. **The same** scale as in Fig. 9, photographed through a 2%
sucrose-water gradient, 2-cm light path, 21 min after forming the boundary.

When one takes into account that the scales were photographed on glass
plates ("for greater accuracy," see Ref. [46]), which were of course transparent, one marvels at the fact that moiré $\partial c/\partial x$, x curves were not acciden-
tally discovered in the 1930's at the latest. It is indeed difficult to understand
how, during uncounted laborious measurements of transparent scales, every-
body involved was able to avoid putting one scale on top of another for several
decades. However, the observation was not made until 1963 and then only as
a result of systematic studies of moiré phenomena [57]. In the meantime the
problem of obtaining $\partial c/\partial x$, x curves quickly and painlessly had been solved
differently (see Sec. V. B).

It must be understood that the dark moiré fringes are formed by the
intersections of the lines of two superimposed scales of gratings. When a
grating is placed upon a slightly coarser grating, it is always possible to do
at such an angle, α, that the resulting moiré fringes become vertical [15].

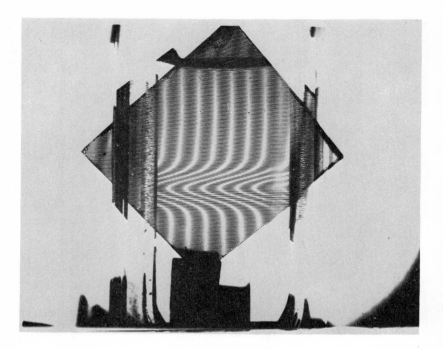

Fig. 11. Moiré formed with photograph in Fig. 9, laid 1.2 mm above photograph in Fig. 10 and interfering with it at an angle of 4.5°. The moiré fringes form a clearly visible series of Gaussian $\partial c/\partial x, x$ curves.

In such a situation, the upward deflection of one of the horizontal lines of the bottom grating over a distance of Δx will cause a sideward displacement of the intersection point of that line with the corresponding line of the top grating of Δy, so that

$$\Delta y = \Delta x \cot \alpha \tag{3}$$

(see Fig. 12).

In this fashion the apparent upward deflection of the multiple horizontal lines placed behind a refractive index gradient can be translated into the sideward development of the $\partial c/\partial x, x$ curves (see Fig. 11). The degree of sideward displacement of the $\partial c/\partial x, x$ curve, or, in other words, the height of the peak, can be changed by changing the angle α between the interfering grating and the rear, horizontal grating. (When changing the angle α, the distance between the two gratings must also be changed, so as to maintain a vertical "base line.") As a general rule, the smaller the angle α, the higher the $\partial c/\partial x, x$ peak; but the thicker the moiré lines and the greater the angle α,

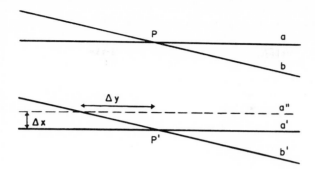

Fig. 12. Two lines, a and a', of a horizontal grating, crossed at P and P' by two lines b and b' of another, equally spaced grating, at an angle α to the first one. The intersection between a' and b' is deviated a distance Δy to the left, when a' is distorted a distance Δx upward to a". Obviously, $\Delta y = \Delta x \cot \alpha$.

the lower the peak but the thinner and sharper the moiré lines. Much the same antithesis exists between the peak height on the one hand and the sharpness of its outlines on the other, as a function of the angle of the sideward deflecting device in schlieren optics (see Sec. V. B).

When the horizontal (distorted) grating is finer than the interfering grating, a moiré $\partial c / \partial x, x$ curve can be obtained by putting the interfering grating some distance <u>behind</u> it (this, of course, holds true for photographs only).

In diffusion measurements it is useful, in practice, to use two fairly coarse gratings for the inspection of the formation of the boundary and for general adjustments of the entire optical arrangement (gratings of the order of 50 lines/in. are quite practical in those circumstances [57]). It is not very practical, of course, to put gratings in water baths, but it is possible to dispense with water baths and just to use water-jacketed cells (Precision Cells Inc., New York, N. Y.), with the help of two circulating thermostats (one kept at $16 \pm 0.1°C$, which serves to cool the second one, which keeps the cell at $20 \pm 0.01°C$ [59]. Once the diffusion is under way, photographs are made through the cell of a horizontal grating of 200 or 150 lines/in. (see Figs. 10, 11). It is useful to take one photograph of the undistorted grating, through the cell filled with pure solvent (Fig. 9). That photograph can then be used as the interfering grating, which minimizes the effects of possible irregularities in the grating itself or in its positioning on the optical bench.

3. Inflection Point Determination by the Scale Method

In cases where diffusion coefficients are to be determined for pure solutes (with diffusion coefficients of a low concentration dependence), it is helpful to be able to locate the inflection points on the $\partial c / \partial x, x$ curve. In such cases there is a very simple connection between the value of x_i at the inflection points and the diffusion coefficient D (see, for instance, [9]):

$$x_i^2 = 2Dt \qquad (4)$$

If it is impossible to establish the time of inception of a sharp boundary with any degree of precision, it may still be permissible [58], after a certain time lapse, to determine D, if x_i has been measured at two different times, by using

$$x_{i_2}^2 - x_{i_1}^2 = 2D(t_2 - t_1) \qquad (5)$$

If a Gaussian $\partial c/\partial x, x$ curve is obtained, it is of course possible to locate x_i as a function of the height of the Gaussian peak (x_{max}) [9] according to

$$\partial c/\partial x_i = \partial c/\partial x_{max} \exp\left(-\tfrac{1}{2}\right) \qquad (6)$$

In other words, one simply situates the inflection points at 60.65% of x_{max}. Unfortunately, however, the exact height of the $\partial c/\partial x, x$ peak is the one most difficult parameter of the curve to measure with any degree of precision, be it with scale, moiré, or schlieren optics.

But the inflection points of the $\partial c/\partial x, x$ curve can also be formed directly, with good precision, simply by means of a photograph of the distorted scale, taken through the gradient. For if, as was shown above, the displacement of each line on the distorted grating is proportional to the local change in concentration, the spacing between the lines of the distorted grating is proportional to the change in the change in concentration $\partial^2 c/\partial x^2$. The maximum and minimum of the $\partial^2 c/\partial x^2, x$ curve then obviously correspond to the place of widest and the place of narrowest spacing between the lines on the distorted grating [15].

These places can be located by checking the distances between the lines of the distorted grating (see Fig. 10). This can be done, for instance, by projecting a photograph of the grating on the screen of a microcomparator and measuring the distances between all the lines with the help of a micrometer screw.

Another, and easier, way is to scan a transparency photograph of the distorted scale with a microdensitometer (Analytrol, Beckman, Fullerton, Calif.), provided with a microzone scanning attachment with a slit width of 0.1 mm. In this manner a 10 X enlarged plot is obtained of the relative optical density of the scanned lines (corresponding to their thickness). When a Ronchi ruling (Edmund Scientific, Barrington, N. J.) is used, which has an equal thickness of black lines and white intervals, the tops as well as the bottoms of the peaks of the plot each form a curve that is homologous to the $\partial^2 c/\partial x^2, x$ curve [59] (their height being proportional to the thickness of the distorted black lines and their depth to the thickness of the equally distorted white intervals). Such a plot, made of the distorted grating in Fig. 10, is shown in Fig. 13. It is clear how readily the maxima and minima of these

Fig. 13. Densitogram of a scale of 150 lines/in., distorted by a 2%
sucrose-water gradient, 21 min after forming the boundary (see Fig. 10),
scanned with an Analytrol densitometer. The tops of the peaks thus obtained
form a curve that is homologous to the second derivative of the concentration
function: $\partial^2 c/\partial x^2, x$.

plots can be discerned. The distance between a maximum and a minimum
(the apparatus and photographic multiplication factor having been taken into
account) equals $2x_i$.

B. Schlieren Methods

The schlieren is without much doubt currently the most used of all methods, not only in analytical ultracentrifugation, but also in diffusion and moving boundary electrophoresis. Its origin is fairly old and its development has
followed relatively tortuous paths. To begin with the direct, but least known,
branch of its development: The first illustration of schlieren (although the
name "schlieren" came much later) was published by Wollaston [3a] in 1800.
He placed an oblique line behind a concentration gradient, upon which "it

appears bent into different forms" and thus "becomes a convenient object for ascertaining the state of any medium under examination." This method was adopted for diffusion studies in 1893 (see Fig. 14) by Wiener [4], who indicated a method for converting the curve graphically into a $\partial c/\partial x, x$ plot with rectangular coordinates (see Fig. 15). Finally, Thovert, in 1914 [10], showed how to transform the skew curve optically into a $\partial c/\partial x, x$ curve with rectangular coordinates, with the help of a cylindrical lens, for all practical purposes giving rise to schlieren optics as we now know them (except for the fact that Wiener [4] and Thovert [10] seemed to be under the impression that the oblique line, or light slit, could only be used under an angle of $45°$ with the x axis).

The other pathway of development started in 1859, when Foucault [42], with the purpose of testing spherical mirrors for possible flaws, blocked off all correctly reflected light rays with a screen so that possible incorrectly reflected rays which bypassed the screen could be observed. Then Töpler [60], in 1867, described very much the same method, by which any irregularity of the lens could not only be shown up, but the precise image of the irregularity could be obtained with great exactitude (this method used a straight-edged opaque screen that could be moved up and down through the focus of a lens, thus obliterating the image of a fine straight light slit or triangle). Töpler named this procedure the "schlieren" method, because it served to show up "schlieren" (German for "streaks") or irregularities in the glasses of lenses. He also applied the method to the study of air waves caused by electric sparks.

The method is still much in use for the study of flames [61], air flow [62], and water flow [63], and is even used in color [64].

But it was Tiselius et al. [65], in 1937, who first adapted the method for visualizing concentration gradients (in the ultracentrifuge), and Philpot [66] who at last, in 1938, combined it with Thovert's [10] principle by using an oblique knife-edge and a cylindrical lens, while Svensson [67] in 1939 again introduced the oblique light slit (see Ref. [10]) and an oblique thread (see Ref. [3a]). The method has since often been alluded to as the Philpot-Svensson schlieren method.

Figure 16 shows the principle of such a schlieren system. It shows how the two undeviated light ribbons are focused at the middle of the oblique slit. The remaining ray goes through the axis of the cylindrical lens and is thus allowed to proceed straight onto the screen. But the light ribbon that is strongly deviated downward hits the oblique slit on one side, so that the one light pencil that can go on has to go through the side of the cylindrical lens and is thus diverted sideways. It is not, of course, important whether the cylindrical lens is vertical or horizontal — that will only determine whether the image will be vertical or horizontal. The angle of the oblique slit or line is important, exactly as in the moiré system: The larger the angle, the sharper the image of the $\partial c/\partial x, x$ curve, but the lower the peak, and vice versa. Particularly for work with the analytical ultracentrifuge, it is useful to know that sedimentation rates of proteins at extremely low concentrations

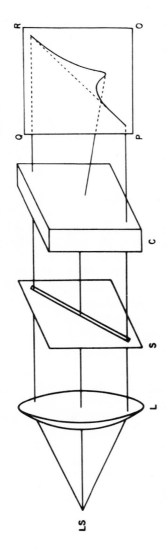

Fig. 14. Wiener's original method [4] for obtaining the $\partial c/\partial x$, x curve with an oblique slit. LS-light source, L-collimating lens, S-slit at 45°, C-cell with gradient, OPQR-screen or photographic plate. Projection, except for the image on the right, which is drawn as an elevation and should be imagined rotated about the line PQ until it forms an angle with the plane of the paper equal to the projection angle used in the drawing.

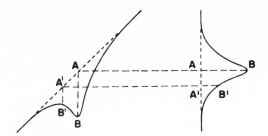

Fig. 15. Method for obtaining the $\partial c/\partial x, x$ curve in a rectangular coordinate system, from Wiener's skew curve. When for every height level
A, A' in the cell, the downward deviation AB, A'B' is measured on the curve
and plotted horizontally vs. the vertical height, the straight $\partial c/\partial x, x$ curve
is obtained. The same effect can be obtained with a cylindrical lens, without
any laborious plotting (see Fig. 16).

(of the order of 0.01%) can still be determined, when angles as low as 10 or
even 5° are used. For that purpose the phase plate (see below) should be
loosened and moved forward in a direction perpendicular to the optical axis
in order to get the schlieren picture back in view [68]. With this technique,
sedimentation coefficients can be obtained at close to zero concentration.
This may in some cases make the practice of determining S-values at several
concentrations, and extrapolating to S_0, superfluous.

The above-mentioned phase plate was the last important improvement in
schlieren optics. As first used by Wolter [69], it was a completely transparent glass disk without any opaque bar or wire, the optical edge being
solely the interruption at the rim of an equally transparent coating covering
exactly half of the glass disk. The coating must be of such a thickness that
the light rays that pass it are one-half wavelength out of phase with the light
rays that pass the uncoated half. A much thinner line could thus be obtained
than was hitherto possible with bars, slits, or wires. The fineness of the
schlieren image could be even further improved by delineating the phase edge
by a very thin wire (4-10 X thinner than the thinnest wire hitherto used), as
first used by Trautman and Burns [70], who used a 75-μ-diameter wire, and
later by Wiedemann, who used a wire as narrow as 20-40 μ [71]. A further
advantage of the use of a phase plate for schlieren optics is that it allows the
simultaneous use of Rayleigh interference optics and will yield both the c,x
and the $\partial c/\partial x, x$ curves in one image [36, 71].

Most of the fundamental work on the theoretical optics of schlieren has
been done by Svensson [72, 73], although a great part of the foundations had
already been laid by Lamm in his work on scale optics [46]. More general
reviews on the subject have been published by Bridgeman and Williams [74]
and by Svensson and Thompson [9]. Recently, Ford and Ford [75] claimed
that in some circumstances important corrections ought to be applied to the
data obtained from schlieren patterns. These claims have been refuted by

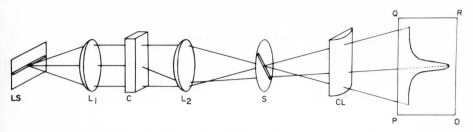

Fig. 16. The schlieren method. This is essentially Wiener's method
(see Fig. 14), with a cylindrical lens which serves to transform the curve into a $\partial c/\partial x$, x curve with right-angle coordinates. LS-horizontal slit with the
light source behind it, L_1 - collimating lens, C-cell with gradient, L_2 -
projecting lens, S-rotatable oblique slit, CL-vertical cylindrical lens, OPQR-
screen or photographic plate. Projection, except for the image on the right,
which is drawn as an elevation and should be imagined rotated about the line
PQ until it forms an angle with the plane of the paper equal to the projection
angle used in the drawing.

Vreeman and Wiersema [76], who pointed out that the problems raised by
Ford and Ford [75] had already been solved 10 years earlier by Svensson
[73]. The apparent difficulty is caused by the finite thickness of the gradient-
containing cell, which unavoidably creates some distortion and tilting of the
schlieren curves (the Wiener skewness [9]). Svensson [73] showed (with
interferometry) how such difficulties can be obviated by focusing on a plane in
the cell situated at one-third of its thickness, counted from the side of the
camera.

VI. CONCLUSIONS

All in all, schlieren is one of the most versatile optical tools in refractometry, and judging by the published literature on analytical ultracentrifugation in the last 15 years, it is by far the most used of all the methods treated
in this chapter.

With diffusion, matters are different: Determinations of diffusion coefficients are done increasingly rarely, so that it is becoming superfluous to
try to decide which optical method is presently used most. The reason for
this state of affairs is that the practice of obtaining molecular weights from
sedimentation coefficients combined with diffusion coefficients [38] is more
and more being superseded by molecular weight determinations by sedimentation equilibrium [77] and approach to equilibrium methods [78-80], notwithstanding the greater hydrodynamic information content of the combined sedimentation (or viscosity) and diffusion methods [38, 81].

With electrophoresis, the situation is also changing: The proliferation of
cheap and simple zone electrophoresis apparatus of various types is quickly

making the moving boundary method obsolete [82]. Zone electropherograms can yield $\partial c/\partial x, x$ curves with the aid of the absorption method, by scanning them in a densitometer. But as the stained strips are often most easily interpreted directly with the naked eye [82, 83], the practice of subjecting them to densitometry is by no means general.

REFERENCES

1.   J. Crank, The Mathematics of Diffusion, Clarendon Press, Oxford, 1964.

2.   A. Fick, Ann. Physik (Leipzig), 170, 59 (1855).

3.   C. Huygens, Treatise on Light, Chap. IV, The Hague, 1690; Dover, New York, 1962.

3a.   W. H. Wollaston, Phil. Trans. Roy. Soc. London, 90, 667, (1800).

4.   O. Wiener, Ann. Phys., 49, 105 (1893).

5.   J. Stefan, Proc. Acad. Sci. Vienna (Math.-Phys. Sec.), 78(2), 957 (1878)

6.   S. H. Armstrong, M. J. E. Budka, K. C. Morrison, and M. Hasson, J. Am. Chem. Soc., 69, 1747 (1947).

7.   G. E. Perlmann and L. G. Longsworth, ibid., 70, 2719 (1948).

8.   R. Barer and S. Tkaczyk, Nature, 173, 821 (1954).

9.   H. Svensson and T. E. Thompson, in Analytical Methods of Protein Chemistry (P. Alexander and R. J. Block, eds.), Vol. 3, Pergamon, New York, 1961, p. 57.

10.   J. Thovert, Ann. Chim. Phys. (9th Ser.), 2, 369 (1914).

11.   R. Zuber, Z. Physik, 79, 280 (1932).

12.   P. P. Debye, J. Appl. Phys., 17, 392 (1946).

13.   G. Kegeles, J. Ann. Chem. Chem. Soc., 69, 1302 (1947).

14.   H. Labhart and H. Staub, Helv. Chim. Acta, 30, 1954 (1947).

15.   C. J. van Oss, J. Sci. Instr., 41, 227 (1964).

16.   T. Young, Phil. Trans. Roy. Soc. London, 92, 387 (1802).

17.   Lord Rayleigh, Proc. Roy. Soc. (London), 59, 198 (1896); 64, 95 (1898).

18.   E. Calvet, Compt. Rend., 220, 597 (1945); 221, 403 (1945).

19.   J. St. L. Philpot and G. H. Cook, Research, 1, 234 (1948).

20.   H. Svensson, Acta. Chem. Scand., 3, 1170 (1949); 4, 399 (1950); 5, 72 (1951); 5, 1301 (1951); 5, 1410 (1951).

21. L. G. Longsworth, Rev. Sci. Instr., 21, 524 (1950); Anal. Chem., 23, 346 (1951).

22. E. Wiedemann, Helv. Chim. Acta, 35, 2314 (1952); Intern. Arch. Allergy, 5, 1 (1954).

23. I. H. Billick and R. J. Bowen, J. Phys. Chem., 69, 4024 (1965).

24. J. C. Jamin, Compt. Rend., 42, 482 (1856); Ann. Chim. Phys., 3, 52, 163 (1858).

25. H. Labhart and H. Staub, Helv. Chim. Acta, 30, 1954 (1947).

26. W. Lotmar, ibid., 32, 1847 (1949).

27. J. W. Beams, A. Robeson, and H. M. Dixon, Rev. Sci. Instr., 25, 295 (1954).

28. J. W. Beams, H. M. Dixon, A. Robeson, and N. Snidow, J. Phys. Chem., 59, 915 (1955).

29. G. L. Gouy, Compt. Rend., 90, 307 (1880).

30. L. G. Longsworth, J. Am. Chem. Soc., 69, 2510 (1947).

31. G. Kegeles and L. J. Gosting, ibid., 69, 2516 (1947).

32. C. A. Coulson, J. T. Cos, A. G. Ogston, and J. St. L. Philpot, Proc. Roy Soc. (London), A192, 382 (1948).

33. L. J. Gosting and M. S. Morris, J. Am. Chem. Soc., 71, 1998 (1949).

34. L. J. Gosting and L. Onsager, ibid., 74, 6066 (1952).

35. A. G. Ogston, Proc. Roy. Soc. (London), A196, 272 (1949).

36. O. M. Griffith and C. R. McEwen, Anal. Biochem., 18, 397 (1967).

37. T. Svedberg and H. Rinde, J. Am. Chem. Soc., 46, 2677 (1924).

38. T. Svedberg and K. O. Pederson, The Ultracentrifuge, Clarendon Press, Oxford, 1940; Johnson Reprint Corp., New York, 1959.

39. V. N. Schumaker and H. K. Schachman, Biochim. Biophys. Acta, 23, 628 (1957).

40. H. K. Schachman, Biochemistry, 2, 887 (1963).

41. H. K. Schachman and S. J. Edelstein, ibid., 5, 2681 (1966).

42. L. Foucault, Ann. Obs. Paris, 5, 197 (1859).

43. O. Lamm, Z. Physik. Chem., A138, 313 (1928); A143, 177 (1929).

44. O. Lamm, Nature, 132, 820 (1933).

45. O. Lamm, Kolloid-Z., 69, 44 (1934).

46. O. Lamm, Nova Acta Reg. Soc. Sci. Upsal. IV, 10, No. 6 (1937).

47.   O. Lamm, Acta Chem. Scand., 9, 546 (1955).

48.   H. Svensson, Nature, 161, 234 (1948).

49.   F. T. Adler and C. H. Blanchard, J. Phys. Colloid Chem., 53, 803 (1949).

50.   O. Lamm and A. Polson, Biochem. J., 30, 528 (1936).

51.   A. Polson, Kolloid-Z., 87, 149 (1939); 88, 51 (1939).

52.   T. Svedberg, Chem. Rev., 20, 81 (1937); Nature, 139, 1051 (1937).

53.   G. Meyerhoff, Makromol. Chem., 5, 161 (1950).

54.   G. Schumacher, Chem.-Ingr.-Technik, 1955, 25.

55.   Y. Nishijima and G. Oster, J. Opt. Soc. Am., 54, 1 (1964).

56.   G. Oster, M. Wasserman, and C. Zwerling, ibid., 54, 169 (1964).

57.   G. Oster and Y. Nishijima, Sci. Am., 208 (5), 54 (1963).

58.   P. F. Mijnlieff and H. A. Vreedenberg, J. Phys. Chem., 70, 2158 (1963).

59.   C. J. van Oss, Proc. Polymer Conf. Series, 2 (1969).

60.   A. Töpler, Ann. Phys. Chem. (Poggendorff's Ann.), 131, 33 (1967); 131, 180 (1967).

61.   B. Lewis and G. von Elbe, Combustion, Flames and Explosions of Gases, Academic, New York, 1951.

62.   W. T. Reid, Schlieren Photography, Kodak Pamphlet P-11, Eastman Kodak Co., Rochester, N. Y. 1960.

63.   J. A. Westphal, Science, 149, 1515 (1965).

64.   D. W. Holder and R. J. North, Nature, 169, 446 (1952).

65.   A. Tiselius, K. O. Pedersen, and I. B. Eriksson-Quensel, ibid., 139, 546 (1937).

66.   J. St. L. Philpot, ibid., 141, 283 (1938).

67.   H. Svensson, Kolloid-Z., 87, 181 (1939).

68.   R. Trautman, J. Phys. Chem., 60, 1211 (1956).

69.   H. Wolter, Ann. Physik, 7, 182 (1950).

70.   R. Trautman and V. W. Burns, Biochim. Biophys. Acta, 14, 26 (1954).

71.   E. Wiedemann, Helv. Chim. Acta, 40, 2074 (1957).

72.   H. Svensson, Kolloid-Z., 90, 141 (1940).

73.   H. Svensson, Opt. Acta, 1, 25, 90 (1954); 3, 164 (1956).

74. W. B. Bridgman and J. W. Williams, Ann. N. Y. Acad. Sci. , 43, 195 (1942).

75. T. F. Ford and E. F. Ford, J. Phys. Chem., 68, 2843, 2849 (1964).

76. H. J. Vreeman and Y. Wiersema, J. Phys. Chem. , 71, 785 (1967).

77. S. Claesson and I. Morning-Claesson, in Analytical Methods in Protein Chemistry (P. Alexander and R. J. Block, eds.), Vol. 3, Pergamon, New York, 1961, p. 121.

78. W. J. Archibald, J. Phys. Colloid Chem. , 51, 1204 (1947).

79. A. Ehrenberg, Acta Chem. Scand. , 11, 1257 (1957).

80. R. Trautman and C. F. Crampton, J. Am. Chem. Soc. , 81, 4036 (1959).

81. H. A. Scheraga and L. Mandelkern, ibid. , 75, 179 (1953).

82. C. J. van Oss, in Progress in Separation and Purification (E. S. Perry, ed.), Vol. 1, Wiley (Interscience), New York, 1968, p. 187.

83. C. J. van Oss, in Advances in Immunogenetics (T. J. Greenwalt, ed.), Lippincott, Philadelphia, 1967, p. 1.

AUTHOR INDEX

Numbers in parentheses are reference numbers and indicate that an author's work is referred to although his name is not cited in the text. Underlined numbers show the page on which the complete reference is listed.

A

Abricassova, I. I. , 127(43), 209
Ackilli, J. A. , 24(63), 39
Adam, N. K. , 1(3), 12(3), 24(57, 59, 66), 36, 39, 59(1), 84
Adamson, A. W. , 138(50), 190(50), 210
Adler, F. T. 225(49), 238
Agin, D. P. , 111(14), 208
Agrawal, J. P. , 100(44), 108
Albertsson, Per-Ake, 48(2), 84
Alexander, A. E. , 8(23, 24, 25), 26(23), 37, 47(42), 86
Ambard, L. , 103(56), 108
Anderson, T. F. , 1(6), 12(6), 37
Andreole, T. E. , 125(38), 180(38), 188(38), 209
Annicolas, D. , 99(35), 107
Annionson, G. , 21(48, 49), 38
Archer, R. J. 44(3), 61(3), 62(3), 67(3), 70(3), 81(3), 82(3), 83(3), 85
Archibald, W. J. , 235(78), 239
Argyle, A. A. , 21(47), 38
Armstrong, S. H. , 216(6), 236
Aylmore, L. A. G. , 45(56), 87
Azuma, K. , 127(45), 210

B

Babakov, A. V. , 127(42), 209
Bach, D. , 50(65), 63(65), 87
Bachrach, H. L. , 97(27), 107
Bakanov, S. P. , 83(34), 86

Bangham, A. D. , 111(15), 208
Bangham, J. A. , 125(38), 180(38), 188(38), 209
Baranayev, M. K. , 43(4, 82, 83), 46(4, 82, 83), 85, 88
Barer, R. , 217(8), 236
Baretta, E. D. , 91(14), 106
Barnes, G. T. , 45(5), 85
Beams, J. W. , 222(27, 28), 237
Bear, R. C. , 210
Bellemans, A. , 9(26a), 37
Benson, A. A. , 111(9), 208
Berry, D. J. O. , 46(48), 71(48), 82(48), 86
Betts, J. J. , 24(61), 39
Beutner, R. , 111(10), 208
Beyrard, N. R. , 104(57), 108
Billick, I. H. , 218(23), 219(23), 237
Bixler, H. J. , 90(1), 99(39), 106, 107
Blanchard, C. H. , 225(49), 238
Blank, M. , 46(18), 47(6, 7, 8, 9, 19), 49(11, 12, 15), 50(64), 54-59(6, 9, 17, 18, 64), 61(19), 63(9, 64), 64(17), 65(6, 18, 19), 71(18), 73(18), 79(8), 80(64), 81(10, 14, 16, 64), 82(7), 83(7, 18), 85, 87
Blatt, W. F. , 91(5), 99(39), 106, 107
Blight, L. , 12(29), 38
Bowen, R. J. , 218(237), 219(237), 237
Boyd, D. P. , 60(20), 85
Bradley, R. S. , 46(21), 85

SUBJECT INDEX

A

Absorption, 57-59, 69
Action potential, 192-197, 207-208
  induced in BLM, 194-197, 207-208
Adsorption, 3, 8-9
  films, 62, 64, 70-71
  rates, 67-70
Animal membranes, 96
Anisotropic membranes, 90-96
Apparatus for BLM, 121-128, 162-
    167, 206
  electrical measurements for,
      162-167, 206
  forming of, 121-128

B

Barcroft apparatus, 73-75
Barriers for film balance, 17-18
Bifacial tension measurements in
    BLM, 187-193
  experimental methods, 190-192
Bimolecular lipid membranes, 49,
    78, 109-211
  historical, 111-113
  theoretical variables, 128-131
Bimolecular lipid membranes in
    single compartment, 121-124
Bimolecular lipid membranes in
    two compartments, 124-128
Binding force measurement, 98-99
Biological membranes, see
    bimolecular lipid membranes
Black lipid membranes, see
    bimolecular lipid membranes
BLM, see bimolecular lipid
    membranes
Brewster angle, 148, 151

Brush method for BLM, 121, 122,
    125, 129, 134
Bubble-blowing method for BLM,
    124, 127, 129, 134

C

Calibration of film balance, 14
Capacitance of BLM, 155-158, 160,
    172-179
  alternating current bridge,
      172-173
  direct current transient, 173-174
  for thickness, 155-158, 160,
      174-179
Capillary membranes, 91, 105
Cellophane membranes, 90, 111
Cellulose acetate membranes, 93-95
Cleaning of glassware for film
    balance, 21
Coagulation of membranes, 92-93
Collodion film membranes, 92, 111
Concentration gradient measurement,
    213-239
Concentration of solute, 99
Convection, 58-60, 72
Cryptoanisotropy, 91

D

Deposition of monolayers for film
    balance, 19-21
  spreading solvent, 20
Desorption from monolayer, 5-8,
    19-21, 34-36
Diffusion, 57, 79-81, 183-186,
    213-215, 225, 229, 235
  barriers, properties of, 79-81
Dipping method for BLM, 121, 129

249